THE ICE-AGE HISTORY OF ALASKAN NATIONAL PARKS

THE ICE-AGE HISTORY OF ALASKAN NATIONAL PARKS

Scott A. Elias

Smithsonian Institution Press
Washington and London

For my parents

Copy edited by Princeton Editorial Associates
Production Editor: Jenelle Walthour
Designer: Alan Carter

Library of Congress Cataloging-in-Publication Data
Elias, Scott A.
The Ice Age history of Alaskan National Parks / Scott A. Elias.
p. cm.
Includes bibliographical references and index.
ISBN 1-56098-423-6 (cloth).—ISBN 1-56098-424-4 (pbk.)
1. Paleo-Indians—Alaska. 2. National parks—Alaska—History. 3. Paleoecology—Alaska—Quaternary.
4. Geology, Stratigraphic—Alaska—Quaternary. 5. Alaska—Antiquities. I. Title.
E78.A3E45 1995
560′.45′09798—dc20 94-26084

Manufactured in the United States of America
01 00 99 98 97 96 95 5 4 3 2 1

The paper used in this publication meets the minimum requirements of the American National Standard for Permanence of Paper for Printed Library Materials Z39.48-1984.

CONTENTS

ACKNOWLEDGMENTS

I thank the National Park Service Publications Office in Denver for financial support that allowed me to get started writing this book. I thank the U.S. Geological Survey Photo Library in Denver for access and permission to use several photographs of geologic features in the parks. I also thank U.S. Geological Survey geologists Tom Ager, Paul Carrara, and Chris Waythomas for providing advice, photographs and other materials, and encouragement. The faculty of the Alaska Quaternary Center, University of Alaska, including James Dixon, David Hopkins, and Dan Mann, provided many useful suggestions and materials. Faculty of the Institute of Arctic and Alpine Research, University of Colorado, including Nelson Caine, John Hollin, Mark Meier, Gifford Miller, Susan Short, Tom Stafford, Mort and Joanne Turner, and Skip Walker made helpful suggestions on first drafts of chapters and provided photographs. The manuscript, or parts of the manuscript, were carefully and thoughtfully reviewed by Phil Brease, Denali National Park and Preserve; Parker Calkin, State University of New York, Buffalo; James Dixon, University of Alaska, Fairbanks; Saxon Sharpe, Desert Research Institute, University of Nevada, Reno; and my acquisitions editor at the Smithsonian Institution Press, Peter Cannell. I thank all of the reviewers for their help and encouragement.

Finally, I thank my family for their support throughout this project and for doing without me through many evenings and weekends, not to mention the many weeks of fieldwork in Alaska through the last decade.

In addition to the support of the National Park Service, other financial support for the preparation of this book was provided by a grant from the National Science Foundation to the University of Colorado for Long-Term Ecological Research, DEB-9211776. National Science Foundation support for my paleoecological research in Alaska has been provided by grants DPP-8314957, DPP-8619310, DPP-8921807, and OPP-9223654.

PART ONE

Paleoecology

Why We Need to Study Past Ecosystems

HAVE YOU EVER WONDERED what the prehistoric world was like? The term "pre-history" encompasses a vast amount of time, and we humans have considerable difficulty conceptualizing the millions and billions of years of Earth's history. We may understand that things haven't always been the way they are now, and we're probably well acquainted with some of the more spectacular episodes in prehistory, such as the age of dinosaurs—if we aren't, our children will be happy to fill us in! But having gotten that far, can we honestly say that we have a good understanding of what a million years means, or what percentage of total Earth history it represents?

At the nearer end of the prehistoric time scale are the events of the last few tens of thousands of years. It was during this time period that the glaciers of the last ice age began to melt, that humans moved into North America from Asia and, more recently, from Europe and Africa. We have come to learn something of that time when nearly all of this continent was clothed in primeval (old-growth) forests or unbroken expanses of tall-grass prairie. In an attempt to preserve some of that untamed wilderness, our government has set aside tracts of land as National Parks, beginning with Yellowstone, in 1872. In some cases, these parks represent the last, best examples of entire **ecosystems** close to their primeval state.

My aim in writing this book is to provide an overview of this more recent period of prehistory for the vast state of Alaska and of the methods used to reconstruct past environments. Drawing from the work of many scientists, I have brought together information from studies of fossil plants and animals and geological data, which can be used to reconstruct ancient climates. Reconstructing interactions between prehistoric plants and animals and their physical environments is called **paleoecology.** Finally, I provide an overview of the early peoples of Alaska, especially the arrival of the first nomadic hunters who came across the **Bering Land Bridge.**

Chronologically, this book covers the late **Quaternary Period:** the last 125,000 years (Fig. I.1). During this interval, ice sheets advanced southward, covering Canada and much of the northern tier of states in the United States. Glaciers also crept down from mountaintops to fill valleys in the Rockies and Sierras. In Alaska, glaciers formed in the mountains in the Alaska and Brooks Ranges, and the ice flowed both north and south of these ranges. The ice from the glaciers on the south side of the Alaska Range came together to form an ice sheet that covered most of south-central Alaska, whereas glaciers on the north side of the range did not extend far beyond the mountain fronts. So, much of the interior of Alaska apparently remained free of ice during the last glacial cycle and, indeed, throughout most of the Quaternary Period.

The late Quaternary interval is important because it bridges the gap between the ice-age world of prehistoric animals (many of which are now extinct) and modern environments and biota. It was a time of great change, both in physical environments and in biological communities. It was also the time when human beings "came of age" in the world; they spread across the continents and started to become a major factor in the world's ecosystems.

The study of late Quaternary paleoecology is necessary to help us fathom modern ecosystems, because modern ecosystems are the direct result of these past events. To try to understand present-day environments without a knowledge of their history since the last ice age would be like trying to reconstruct the plot of a long novel by reading only the last page.

You will meet few unfamiliar plants and animals in this book. With occasional notable exceptions, they are still growing or cavorting around on the North American landscape (albeit some more slowly than others). Those species that have become extinct since the ice age are a fascinating story unto themselves, one that we shall explore. Fossil studies show that the modern ecosystems did not spring full-blown onto the hills and valleys of our continent within the last few centuries. Rather, they are the product of that massive reshuffling of species that was brought about by the last ice age and, indeed, continues to this day.

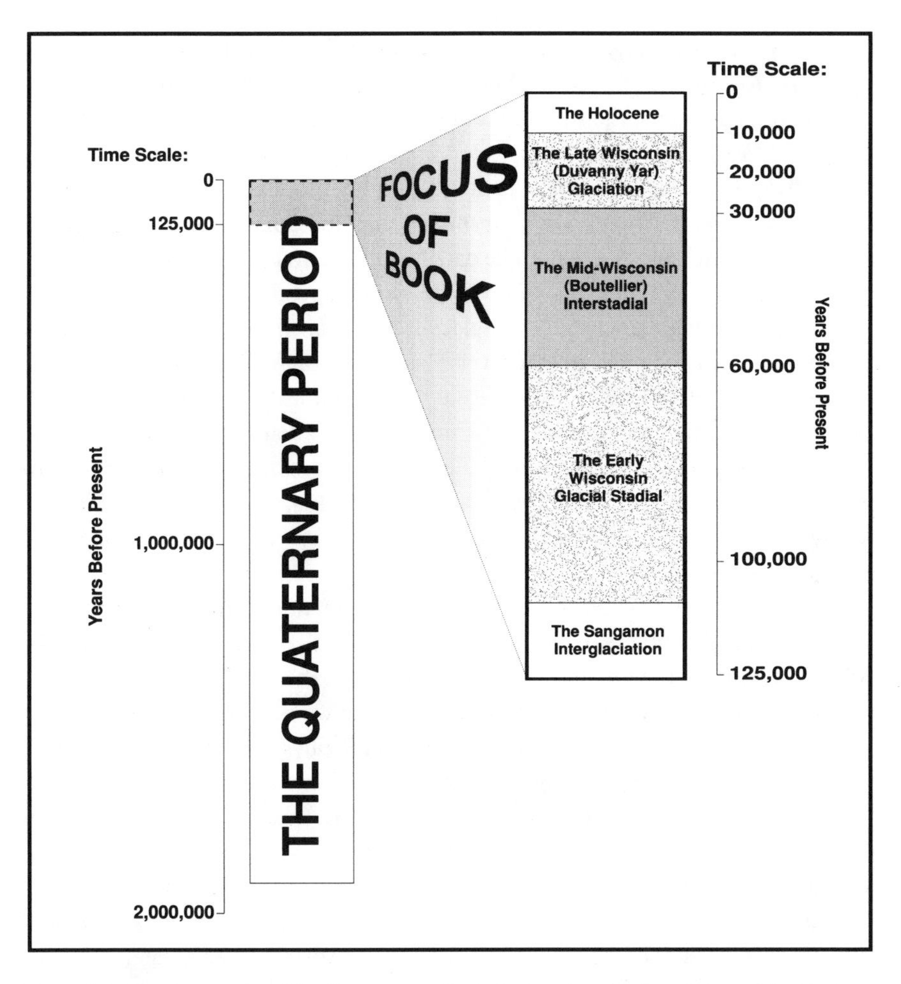

Figure I.1. The time scale of the Quaternary Period, showing the time interval that is the focus of this book: the late Quaternary, spanning the last 125,000 years.

The study of past ecosystems is really just a form of detective work. A police detective reconstructing a crime has to reconstruct the following aspects of the case:

1. What happened?
2. Who did it?
3. How was it done?
4. When was it done?
5. What was the motivation for the crime?

Apart from the last category, paleoecologists are basically saddled with the same questions. You might say that in fossil studies, the trail of clues has grown extremely cold. Certainly, the suspects and witnesses are all long since dead. Nevertheless, the task, although difficult, is not impossible. Moreover, it's often exciting and challenging. And, as we shall see, it is becoming more important all the time.

In order to reconstruct *what happened* in the Quaternary, we need to look at both physical and biological data. The data from the physical environment include such information as types of sediments deposited and their rates of deposition, types of landforms associated with glacial and near-glacial (called *periglacial*) environments, and changes in lake levels. The fossils themselves tell us *who did it*, although we are often looking for external factors that influence the environment, such as changes in the amount of incoming solar radiation (called *insolation*). Other aspects of the physical environment make occasional appearances on the "who done it" list. These include volcanoes, earthquakes, and even the odd meteor or asteroid. The dating methods outlined in Chapter 3 provide information on *when it was done.*

Perhaps the trickiest question of all is, *"How was it done?"* In other words, how have the various elements of the physical and biological world interacted on the global stage during the Quaternary? This question is the most difficult to answer, but it is also perhaps the most important, because we desperately need to know how the biological world responds to changes in the physical environment. We are inflicting our own changes on the environment of planet Earth at an ever-increasing rate. Overhunting, overfishing, pollution, and destruction of natural habitats have already wrought havoc on most ecosystems and have caused the extinction of untold numbers of species within the last few centuries. At present we find ourselves in the position of trying to understand how to sustain the remaining flora and fauna just as human activities place the natural world in ever-greater jeopardy. Virtually our only means of gaining a greater understanding of how regional biotas respond to environmental change is to examine how these biota have responded to past changes. The current crisis takes paleoecology out of the arcane realm of satisfying the intellectual curiosity of a few eccentric college professors and places it squarely in the middle of worldwide efforts to save our remaining biota.

In this volume, I am focusing on the prehistory of four National Parks and Preserves in Alaska (Fig. I.2), for two reasons. One is that, as nature preserves, the parks offer excellent opportunities for scientific research in many fields, including paleontology and geology. Because the parks are relatively pristine, it is easier to compare past and present ecosystems in them than in regions that have been greatly modified by human activities. Another reason for my focus is that many people are interested in these parks and in the ecosystems they represent. This

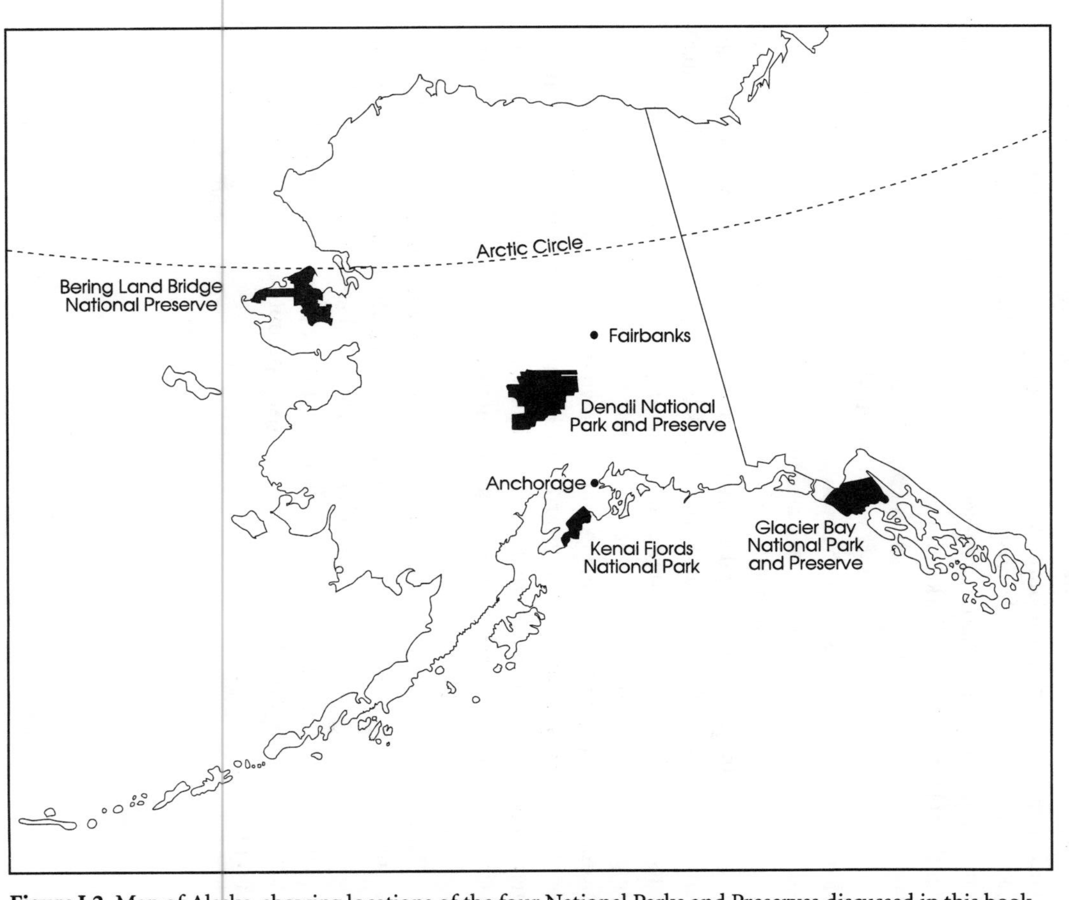

Figure I.2. Map of Alaska, showing locations of the four National Parks and Preserves discussed in this book.

book is an attempt to fill the gap between books available about modern ecosystems and those concerning bedrock geology.

Obviously, there is a vast gap between the events of millions of years ago and the events of the last hundred years. The first step in bridging that gap is providing information on the hows and whys of paleoecology. Then we will examine the ancient environments, vegetation, animals, and peoples of the Alaskan interior (at Denali National Park and Preserve), the Bering Land Bridge and western Alaska (at Bering Land Bridge National Preserve), the rugged coast of the Kenai Peninsula (at Kenai Fjords National Park), and the ever-shifting environments of Glacier Bay, where the ice age is still very much in progress (at Glacier Bay National Park).

Let me issue one caution before we go any further. If you are reading this book, it is probably safe to assume that you already have some interest in fossils or archaeological artifacts. If you visit a national park and spot either fossils or artifacts, please do not touch them. The best thing to do is to mark the spot, or leave someone at the spot, and go find a park ranger. Tempting as it might be to pick up an arrowhead or a fossil bone and bring it back to a ranger station or visitor center, this would be the wrong thing to do. By removing a fossil or artifact from its original location, you would be destroying the evidence of exactly where it came from. This evidence is often vital in pinning down the age, environmental context, or cultural context of fossils and artifacts. Fossils and artifacts that have been removed from the places where they were originally deposited cease to be useful to science; they become mere souvenirs of ancient times, stripped of the context that gives them much of their meaning.

Speaking of souvenirs, of course, it is illegal to remove fossils and archaeological materials from national parks (along with rocks, soils, plants, and animals, except fish caught with a park-issued license). Furthermore, all research done by qualified scientists in national parks is carried out only with the permission of the National Park Service. Science officers in the parks issue permits for specific projects at specific locations.

1

QUATERNARY FOSSILS

What Are They, and Where Are They Found?

Simply put, Quaternary fossils are the remains of organisms that lived during the Quaternary Period. The Quaternary Period is the most recent of the geologic periods; it is the time interval covering the last 1.7 million years and is characterized by numerous glaciations. At the turn of the century, geologists thought there had been four major glaciations in the Quaternary in North America. Now it appears that North America experienced at least 14 Quaternary glaciations. These glacial intervals are lumped together into the **Pleistocene epoch,** which began with the onset of the first glaciation (1.7 million years ago) and ended with the last glaciation. The last 10,000 years (the interval since the end of the last ice age) are called the **Holocene.** This book covers events occurring at the very end of the Pleistocene and during most of the Holocene.

All of written human history has taken place in the second half of the Holocene. The last Pleistocene glaciation in North America, called the **Wisconsin Glaciation,** was probably the most important force affecting the development of our modern ecosystems; it had an effect on every part of the continent, even though it did not cover the whole continent with ice. Subsequent retreat of the glacial ice opened the way for recolonization of continental landscapes and brought a massive reshuffling of species in every region of North America.

Vertebrate Fossils

Several types of Quaternary fossils are commonly studied by paleontologists. Vertebrate fossils often dominate paleontological literature, museum displays, and popular literature. This notoriety is based on their high visibility and our own affinity for mammals. The skeleton of a woolly mammoth has the capacity to capture our imagination, whereas most of us have a hard time getting excited about pollen grains, diatoms, or bits and pieces of insects. When people think of ice-age fossils, they usually remember the fossil bones of large, extinct mammals, such as mammoths, mastodons, and saber-toothed cats, that they have seen displayed in museums. Although these fossils are fascinating and informative, they are actually quite rare in comparison with the fossils of smaller animals and plants.

It is sometimes difficult to make paleoclimatic interpretations based on the fossil remains of large animals, because many of them probably migrated across landscapes on a regular basis and thus were able to avoid undesirable climatic conditions. On the other hand, small mammals, such as rodents, shrews, rabbits, and bats, not to mention fish, birds, reptiles, and amphibians, offer tremendous opportunities for paleoenvironmental reconstructions because they generally remained in their small home range year-round. For instance, fossils of the collared lemming, still an inhabitant of arctic tundra, have been found in late Pleistocene deposits in the American Midwest. These data provide convincing evidence that the climate of central Iowa in the late Pleistocene was quite similar to the climate found today on the Alaskan North Slope.

Although the vertebrate fossil record is important, scientists have gathered far more information from other kinds of fossils, most of which are much less glamorous; in fact, most are practically invisible to the naked eye. Among these are the pollen, stems, leaves, and fruits of plants, the **exoskeletons** of insects, the shells of snails and other mollusks, and the glassy skeletons of microscopic algae, such as **diatoms.** These types of fossils are small, but they are much more abundant in sediments than the bones of ancient mammals. In fact, just a thimbleful of lake sediment may contain thousands of pollen grains or diatoms, all wonderfully preserved down to the last detail of **microsculpture,** as viewed through a high-powered microscope. By compiling the data gathered from all of the various types of fossils from a given time period in a study region, teams of scientists are able to piece together a picture of its plant and animal life. They can then use the paleontological data to reconstruct the history of climate change. We will now focus on the major types of **microfossils** used in Quaternary research.

Fossils from Plants: Pollen and Macrofossils

Palynology is the study of pollen. It is probably the most widely used tool in terrestrial Quaternary paleoecology. Many kinds of plants produce a superabundance of pollen each year. This is especially true of wind-pollinated plants, such as conifers (evergreens). A single lodgepole pine may produce as much as 21 billion pollen grains per year. Other plants, such as insect-pollinated species, may produce only a few thousand grains per year. Pollen has an extremely durable outer wall, which resists decay. The pollen grains of many plant species are light enough to float on the wind and may travel hundreds or even thousands of miles before landing. If they land in a lake, pond, or bog, they may be preserved for thousands of years.

Pollen samples are usually extracted from sediment cores taken from lakes and ponds; pollen is also preserved in buried soils, in peat, and even in glacial ice (the Greenland Ice Cap has been found to contain the pollen of spruce trees that grew thousands of miles away, in Canada).

Differences among the pollen grains of different species in a genus are often difficult or impossible to detect under a light microscope. Because of this, pollen grains are most often identified only to the family or genus level rather than to the species (see Fig. 1.1 for illustration of taxonomic hierarchy). For instance, a pollen grain may be identified as coming from a pine tree, but it is not always possible to tell which species of pine (such as white pine, jack pine, or lodgepole pine) the pollen comes from.

Once the pollen grains are identified, the data are generally presented as diagrams showing pollen percentages of the total number of grains in sediment samples, plotted according to depth in a **stratigraphic column** (Fig. 1.2). Pollen diagrams from the same region and same time interval are generally quite similar to each other but are different from diagrams from other regions or times. Pollen diagrams are often divided into zones of similar pollen composition. The boundaries between zones mark transitions in regional vegetation. The proportions of pollen released into the environment depend on the numbers and types of plants and therefore reflect the composition of regional vegetation. For instance, following the retreat of the Wisconsin Ice Sheets in North America, some regions were first colonized by low, herbaceous vegetation (grasses, sedges, and their relatives), similar to what is now found in the arctic tundra. As regional climates warmed and soils began to mature, the herbaceous vegetation gave way to coniferous forest, which in turn was invaded by hardwood trees, eventually producing the mixed deciduous–coniferous forests of today. The timing of these transitions in plant communities is usually clearly shown in regional pollen diagrams.

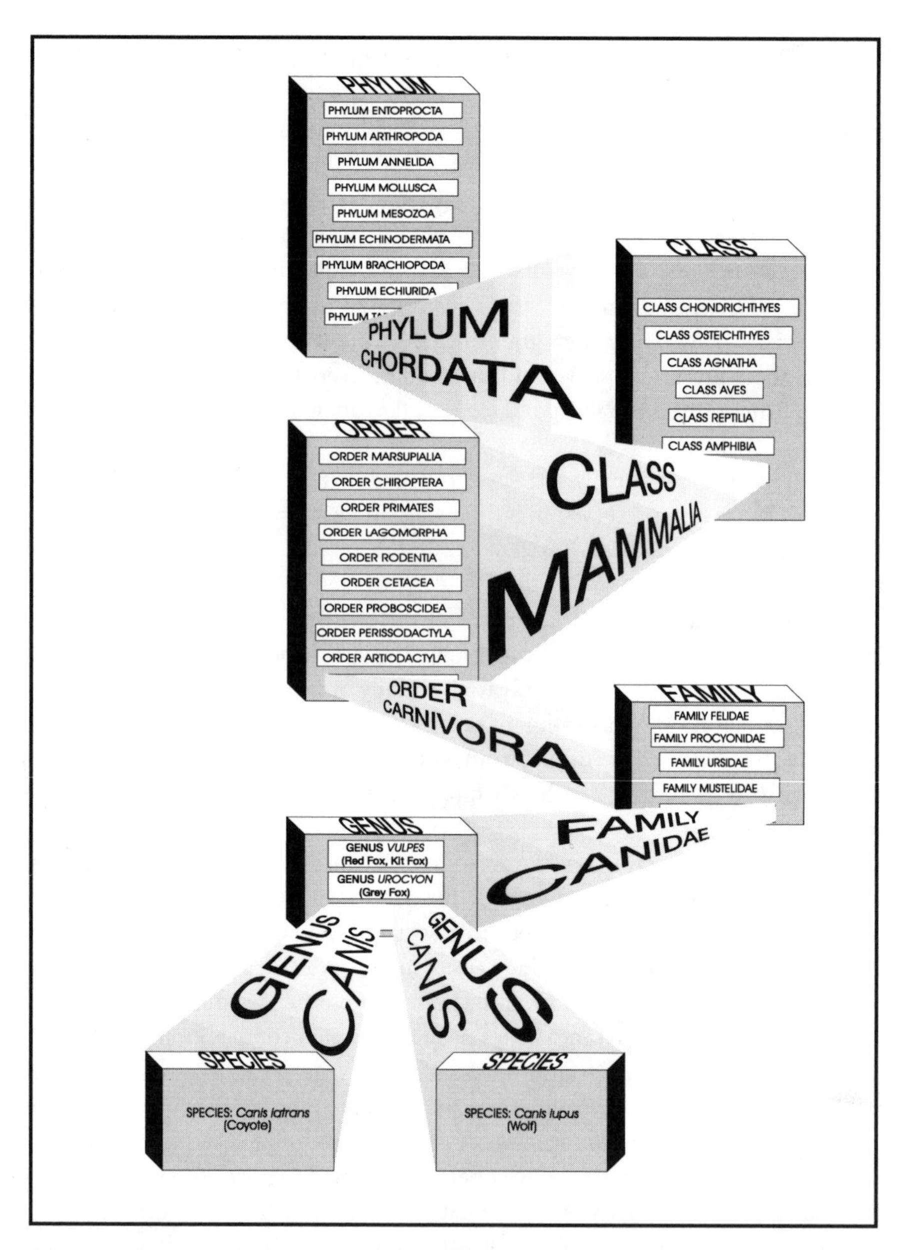

Figure 1.1. Illustration of the hierarchy of classification of the animal kingdom, using the example of the wolf and coyote, two species in the genus *Canis,* which is one of several genera in the family Canidae (dogs, wolves, coyotes, and foxes). The canids represent one family in the Order Carnivora (meat-eating mammals). The carnivores, in turn, are one order in the Class Mammalia (mammals). The mammals are one class in the Phylum Chordata (animals with spinal cords). There are many phyla in the Kingdom Animalia (the animal kingdom).

Recently, palynologists have developed statistical methods of reconstructing past environmental conditions by comparing the pollen "signature" of modern stands of vegetation with fossil pollen assemblages. By analyzing the climatic conditions under which the modern vegetation is growing and the proportions and amounts of pollen types in modern sediments from the various types of plant communities, they have been able to come up with fairly precise estimates of past climates. In addition, palynologists have used pollen diagrams to track the long-term movements of plant species and genera through the late Quaternary. Long-lived plants, such as most trees, do not race across landscapes. Their migration in response to climate change may take centuries, but they have undergone some remarkable shifts in distribution through the past few thousand years. This may seem like a long time to us, but it represents only a small fraction of the time that these plant species have been in existence.

The macroscopic (visible to the naked eye) remains of plants commonly preserved in Quaternary sediments are called **macrofossils.** Woody plants produce a large number of potential macrofossils through their life cycle. As might be expected, wood is very resistant to decay in most waterlogged sediments. Specialists in fossil wood identification are able to determine species of trees on the basis of analysis of cross sections of stems. The annual growth rings in trees can provide

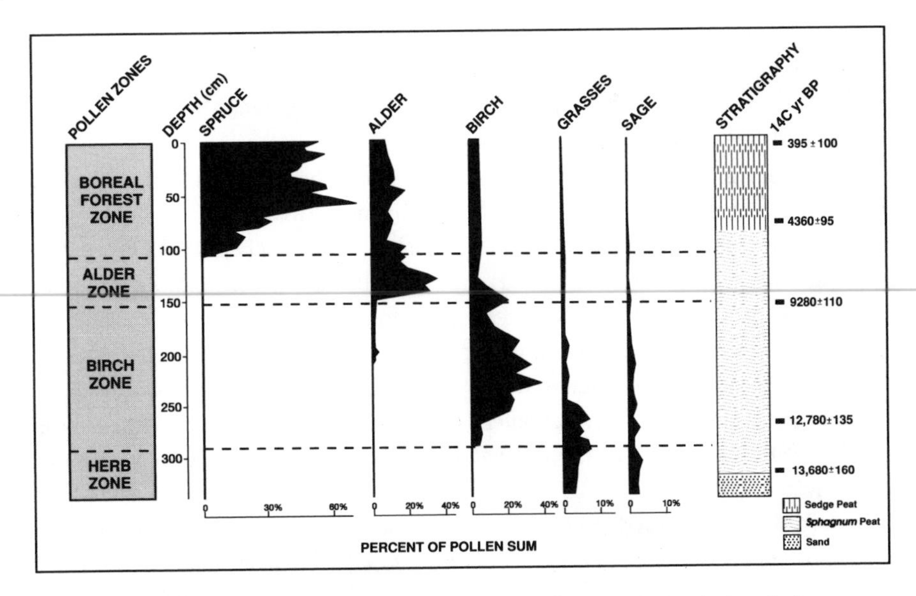

Figure 1.2. Generalized percentage pollen diagram for sites in Alaska during the last 14,000 years. Note that the pollen zone boundaries reflect the major changes in pollen percentages through time.

both a history of local environments (for example, drought versus abundant precipitation, or episodes of forest fires or insect infestation) and a year-by-year chronology of those events. Macrofossils from coniferous trees also include needles and cones, both of which can frequently be identified to the species level.

Nonwoody plants, such as grasses and herbs, also produce a wide variety of macrofossils, including stems, leaves, and fruits. Peat is essentially composed of layer upon layer of undecomposed plant leaves and stems. The two principal types of peats are moss peats (frequently dominated by sphagnum mosses) and sedge peats. Moss fragments can often be specifically identified. Sphagnum mosses from bogs dominated many regions during the Pleistocene, and the combination of living and dead (peat) mosses probably represents the largest single terrestrial repository of carbon.

Plant macrofossils generally accumulate in **water-lain sediment** and reflect only local conditions, since they come from plants growing in the **catchment basin.** In order to develop a regional reconstruction of past vegetation based on macrofossils, it is necessary to study multiple sites and piece together the data from each site into a synthesis of information.

Insect Fossils

Insect fossil studies began in earnest during the 1960s and have become one of the most important sources of terrestrial data on past environments. These studies have, for the most part, employed fossil insects as **proxy data,** that is, as indirect evidence for past environmental conditions.

Beetles are the largest order of insects. They have been the main insect group studied from Quaternary sediments. They are the most diverse group of organisms on Earth, with more than one million species known to science (that's more than all of the flowering plants combined). In addition, their exoskeletons, reinforced with **chitin,** are extremely robust and are commonly preserved in large numbers in lake sediments, peats, and some other types of deposits. In most cases, beetles have quite specialized habitats that apparently have not changed appreciably during the Quaternary. This makes them excellent environmental indicators. The exoskeletons of beetles and some other insects are covered with exquisite microsculpture, enabling paleontologists to identify fossil exoskeletons to the species level in at least half of all preserved specimens, even though insect exoskeletons are most often broken up into their individual plates in fossil specimens.

Beetles are very quick to colonize a region when suitable habitats become available. They often respond more quickly than plants, which, until recently, were relied upon almost exclusively as indicators of environmental change on land. Like plant macrofossils, insect fossils are generally deposited in the catchment basin in which the specimens lived. Thus, they provide a record of local conditions, in contrast to pollen, which can be carried many miles on winds and often gives a more regional "signal."

Studies of insect fossils in two-million-year-old deposits from the high arctic have failed to show any significant evidence of either species evolution or extinction. Beetle species have apparently remained constant for as many as several million generations.

Insect fossils are generally extracted from organic-rich lake or pond sediments or peats. Ancient stream flotsam, deposited in **fluvial sediments** and later exposed along stream banks, is often a rich source of insect fossils.

Insect fossil data are usually presented in minimum numbers of individuals for each species identified. Paleoclimatic reconstructions are generally made on the basis of the climatic conditions in the region where the species in a given assemblage can be found living together today, that is, the climate of the region where their modern distributions overlap.

Diatoms, Ostracodes, and Others

These groups of organisms can provide information on ancient lake conditions. This information, in turn, can often be tied directly to climate, because the size and water quality of lakes are controlled largely by the balance among evaporation, precipitation, and local ground-water conditions. The fossil organisms in lakes reflect changes in regional climate, as those changes affect lake water temperature, salinity, alkalinity, and other factors. However, sometimes local factors, such as the presence of springs, alkaline soils, or proximity to glacial ice, may overwhelm the regional climatic signal.

The group of primitive plants lumped together under the common name "algae" is quite diverse and extremely successful, having colonized most regions of the globe, including the land, fresh water, and the seas. The most diverse group of algae are the diatoms. They also happen to be the best-preserved algal group in Quaternary sediments because they produce cell walls made of silica, which resists decomposition. Because diatoms are preserved well and grow in large numbers in both lakes and the oceans, their siliceous remains can reach concentrations of one billion per cubic centimeter of sediment!

Diatoms are not often used for paleoclimatic reconstructions because their presence in a body of water is more dependent on water quality than on regional climate. Nevertheless, they can be quite useful in revealing a number of ancient conditions. For instance, some freshwater diatoms live only in shallow water, whereas others live only in deep water. Some of those living in shallow water cling only to certain types of substrates (rocks, sand grains, mud, or plants). Other important factors controlling the type of diatoms found in a lake include the salinity, nutrient level, and acidity or alkalinity of the water.

There are a host of other "critters" in lakes and ponds that leave behind fossils that can serve in paleoenvironmental reconstructions. Among these are freshwater sponges, water fleas, snails (both terrestrial and aquatic), and ostracods. Freshwater sponges filter water through their pores, straining out bacteria for food. This makes them delicate monitors of water quality. The "skeleton" of sponges is made up of silica spicules, which preserve well in lake sediments and can often be identified to the species level.

Water fleas, or cladocerans, are one of the most important groups of tiny crustaceans that live in fresh water. Like diatoms, water fleas are divided into groups that live in shallow and deep waters. They, too, are seldom used in paleoclimatic reconstructions, but their fossils can tell a great deal about past lake conditions, including nutrient levels, salinity, and acidity/alkalinity. The exoskeleton of water fleas is made of chitin.

Ostracods are tiny, bivalved crustaceans that live in fresh or salt water. Their shell, or carapace, is reinforced with calcite (crystalline calcium carbonate), which dissolves in nonalkaline waters. Their fossil shells are therefore found in nonacidic sediments, such as **marls.** The abundance and diversity of ostracods in a given lake are usually dependent on salinity, degree of oxygenation, acidity/alkalinity, water depth, and food availability. Some are found only in lakes; others prefer ponds; still others live in running water.

A variety of these aquatic plants and animals can be found in abundance in water-lain sediments. It may seem redundant to study more than one or two types of fossils, since many of them provide overlapping information on such parameters as water quality and water depth. However, paleoecologists don't look at it that way. They consider each piece of fossil data to be relevant, because each fossil group usually has some unique data to contribute, and the combined information always makes for a better, more sharply defined picture of past environments. If nothing else, different data sets serve as corroborative evidence, confirming the information provided by some of the more commonly studied groups. Paleoecology is basically detective work, with fossils as witnesses to past environments. The fossils have a lot to say if you can understand their language. The process of paleoenvironmental reconstruction is a lot like fitting together pieces of a jigsaw puzzle. The more pieces you have, the more successful you'll be (Fig. 1.3).

One final word about vertebrate fossils: often the bones of large, conspicuous animals tell less about past environments than the bones of very small, inconspicuous animals, such as rodents. There are several reasons for this. One is that large

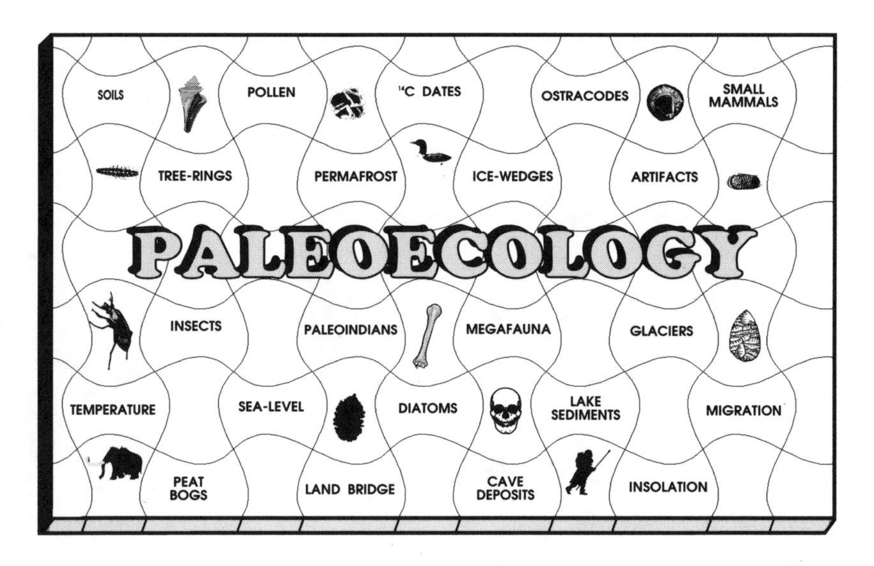

Figure 1.3. In some ways, paleoecological reconstruction resembles the piecing together of a jigsaw puzzle.

mammals are rarer than small mammals in any given ecosystem (for instance, there were many more mice than mammoths on the Beringian landscape during the Pleistocene). Consequently, a given fossil assemblage may produce thousands of bones (or pieces of bones) from voles, mice, and squirrels but only a few bones of large mammals. Second, paleoecologists believe that large mammals provide less useful data about past environments because they are able to move about the landscape and avoid undesirable conditions. On the other hand, small mammals tend to be much less mobile; at least, they do not migrate across whole regions. Because of this, their fossil remains provide a more reliable indication of the environmental conditions at or near the place where their fossils are found.

Suggested Reading

Birks, H. J. B., and Birks, H. H. 1980. *Quaternary Palaeoecology.* London: Edward Arnold Publishers, Ltd. 289 pp.

Eicher, D. L. 1976. *Geologic Time.* Englewood Cliffs, New Jersey: Prentice Hall. 150 pp.

Elias, S. A. 1994. *Quaternary Insects and Their Environments.* Washington, D.C.: Smithsonian Institution Press. 284 pp.

Warner, B. G. (ed.). 1990. *Methods in Quaternary Ecology.* Geoscience Canada Reprint Series No. 5. St. John's, Newfoundland: Geological Association of Canada. 170 pp.

2

THE REPOSITORIES OF ECOLOGICAL HISTORY

How Are Ice-age Fossils Preserved?

When I mention to people that I work on insect fossils, they often assume that I have to hack away at outcrops of shale, looking for faint impressions of fossils in the stone. In fact, Quaternary fossils are most often found in mud, not in stone. That is because most Quaternary sediments are **unconsolidated;** that is, they are loose soils, sands, clays, and other matter that have not yet turned to stone (given a few million years more under pressure and/or great heat, they may). Quaternary fossils can also be found in peat moss (and you thought it was only good for gardening!). Peat is composed of waterlogged plants that accumulate in bogs because decomposition there is extremely slow. Given a few million years under the right conditions, peat turns to coal.

Pre-Quaternary fossils in bedrock generally fall into one of two categories: they are either mineral replacements of the original organic matter or impressions of ancient plants and animals. In both cases, the original organic matter has long since decomposed. However, most Quaternary fossils are the actual remains of plants and animals that have been preserved before decomposition set in. As mentioned above, this type of preservation occurs in peat bogs. It also takes place at the bottom of lakes and ponds, where the sediments are poor or lacking in oxygen (a prime requirement for most types of decomposition). In very dry environments, such as desert caves and rockshelters, the remains of plants and animals gradually dry out and may become "mummified" fossils that last tens of thousands of years.

Quaternary fossils are found in a wide variety of sediments. We will consider only terrestrial and freshwater sediments in this chapter. These materials are quite familiar to everyone. They are the soil in your garden, the smelly mud at the bottom of a pond, and the dust blowing into your house during a windstorm.

Types of Sediments Containing Fossils

Mineral sediments, such as gravel, sand, silt, and clay, accumulate in four main types of deposits: eolian, alluvial, colluvial, and lacustrine. Pleistocene glaciers and ice sheets created vast quantities of sediments as they pulverized the rock over which they moved. Glaciers act as giant conveyor belts, transporting rocks and finer debris both on top of and inside the layers of moving ice. Glacial deposits range from huge boulders to fine silts and clays that get left behind when the ice retreats.

Eolian Deposits

Eolian deposits are wind-blown silts or sands. Loess is a deposit of relatively uniform, fine soil, mostly silt, that was transported to the deposition site by wind. Most loess deposits are relatively inorganic and contain few fossils. Loess mantles large regions of North America, including Alaska, Washington, and Nebraska, in depths up to several tens of meters.

Layers rich in organic matter within permanently frozen silts are an important source of Quaternary fossils in Alaska. They include mammal carcasses that were frozen into the "muck" during the Pleistocene and that have essentially been freeze-dried in the permafrost. This remarkable process has produced several large mammal specimens more or less intact, with all their soft tissues dried to the consistency of very old beef jerky.

Alluvial Deposits

Alluvial or stream deposits are created by moving water; they include cobbles, gravels, sands, and silts. Stream deposits often include pockets of organic debris that accumulated in backwaters, pools, or low-energy side channels. These deposits can be a treasure trove of fossils because they consist of stream flotsam, which may accumulate in large quantities over a short time. When river channels shift, they sometimes abandon a side channel, which then becomes an oxbow lake. Once an oxbow lake forms, it rapidly fills in with vegetation and eventually becomes dry land. In the process, however, the water-lain organic detritus in the oxbow becomes part of the fossil record.

Colluvial Deposits

Colluvial deposits are created by the downslope movement of sediments, often by slope wash or mudflows during storms. In Alaska, saturated soils creep downslope over frozen ground or are frost-heaved downslope in processes called *solifluction.* Solifluction is a cold-region phenomenon that was more widespread during the Pleistocene than it is now. Slope wash, mudflows, and solifluction transport organic materials that lie on or near the surface to the bottom of a slope. With slope wash, this takes place in the minutes or hours during and after a storm; with mudflows, it may stretch to days or weeks. With solifluction, such deposition takes decades to centuries. However, the ultimate results of the transports are similar, namely, the movement of organic materials from upland sediments down to streams, ponds, and lakes, where they are redeposited.

Organic deposits are also contained in buried soils. These occasionally include usable fossils, notably pollen. However, most organic matter in soils is more or less decomposed, so fossil preservation is generally not as good as in water-lain sediments. The other problem with buried soils is that the process of soil development does little to concentrate fossils, so they are few and far between.

Lacustrine Deposits

Water tends to concentrate organic detritus (such as stream flotsam and organic debris in lakes) into recognizable layers that eventually produce more specimens per cubic centimeter than neighboring upland soils. Therefore, the principal types of sediments containing abundant fossils are those that are laid down in water (both standing and running). Water in a basin, such as a pond or lake, collects large amounts of plant and animal matter from its watershed (Fig. 2.1). Dead insects, leaves, seeds, and twigs are washed in by streams. Insects, pollen, and small seeds are blown in by the wind or washed in by streams. Some animals (both large and small) simply fall into the water and drown or are washed downstream after they die.

Taphonomy: How Fossils Become Preserved

The process by which a living organism becomes preserved in the fossil record is called **taphonomy.** Paleontologists have devoted entire careers to the study of

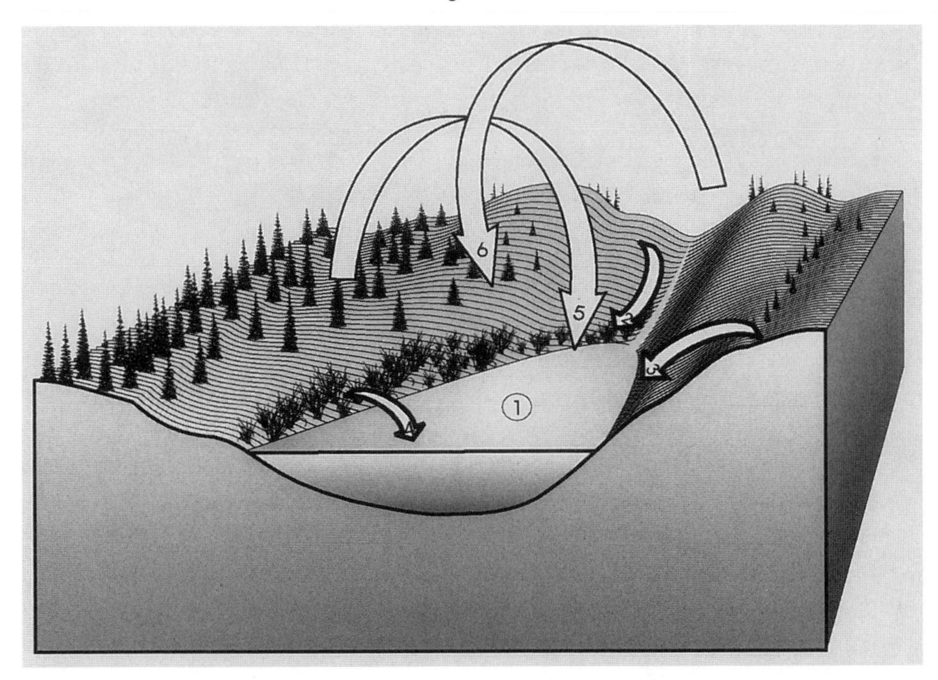

Figure 2.1. Generalized drawing of a mountain watershed, showing sources of potential fossils that may be deposited in a lake: 1, plants and animals that live in the lake; 2, plants and animals carried into the lake by a stream (including stream-dwellers and organisms that fall into the stream); 3, plants and animals transported to the lake by slope wash, erosion, or solifluction; 4, lakeshore plants and animals that fall into the water; 5, microfossils (mostly pollen and spores) from nearby plant communities, transported to the lake by wind; 6, microfossils (mostly pollen and spores) from distant plant communities, transported to the lake by wind.

taphonomy, or "the laws of burial." Figure 2.2 presents a brief summary of the taphonomic process for a lake or pond in a catchment basin. Although the figure is an oversimplification of a very complex set of events, it shows the basic sources of potential fossil material on a living landscape and the general processes that lead to the preservation of plants and animals in sediments.

Most organisms that end up in the bottoms of lakes and ponds spent their lives either in the water or in close proximity to it. In fact, it makes intuitive sense that more aquatic and riparian (shore-dwelling) organisms are preserved in water-lain sediments than upland species, simply because the odds are smaller that an upland creature will make its way into the lake. Some aquatic invertebrates (caddisfly larvae and ostracods, for example) go through several stages of development before reaching maturity. With each new stage, they shed their old exoskeleton or

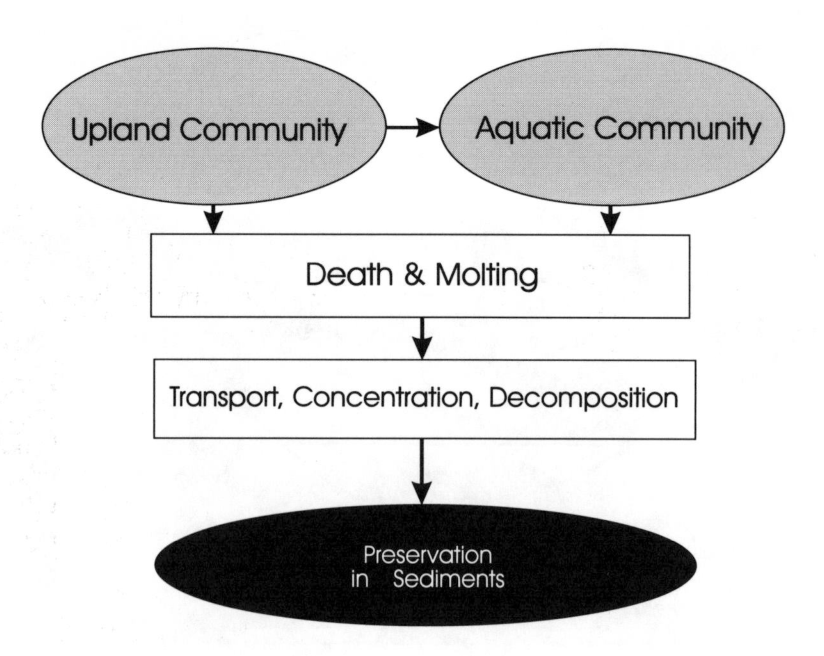

Figure 2.2. Summary of the taphonomy of fossils in a catchment basin.

shell. This molting process creates several sets of potential fossils for each individual.

Water-lain sediments do not preserve a uniform record of regional plants and animals. The remains of many upland creatures, both plants and animals, decompose on the surface and are not preserved as fossils. Even the boniest, most hard-bodied creatures will decompose if left to rot on a hillside. Some large upland animals die near or in the water or are washed downslope by rains, so their carcasses end up in water-lain sediments, but on the whole, we have a better understanding of what life was like in or near the water and a poorer knowledge of ancient life on dry hillsides.

Microfossils, including many types of pollen grains, float through the air and travel quite readily for many miles. This ensures that the pollen that rains out of the sky is representative of a broad region, including uplands and lowlands. Once again, however, pollen and macrofossils from aquatic or shoreline vegetation are often overly abundant in water-lain sediments.

Similarly, upland insects land in lakes and streams by accidentally falling in or by being blown into the water during flight. The proportion of upland and aquatic or shore-dwelling insects in most fossil assemblages is surprisingly well balanced.

Vertebrate skeletons (except for fish) are only occasionally preserved in water-lain sediments. Most good vertebrate fossil deposits have been found in caves or other natural traps. The taphonomy of vertebrate fossils is summarized in Figure 2.3. Bones may be broken or cracked when the animal dies (either by the impact of a fall into a natural trap or by predators); shortly after death, they may be affected by gnawing or breakage by scavengers or by trampling of other animals. Once a carcass is reduced to an accumulation of bones, chemical and physical weathering may come into play, and such factors as frost-heaving, soil erosion, mudflows, and other natural phenomena may move the bones around and mix them with other bones in a deposit.

Let's look more closely at the types of deposits that serve as repositories for Quaternary fossils. To the paleontologist, these repositories represent a bank vault, ready and waiting to offer up a treasure trove of fascinating clues to the history of life on this planet.

Lake Sediments

Lacustrine deposits, including pond and lake sediments, are among the best sources of many types of Quaternary fossils. In addition, a great deal of information about past environments can be gleaned from the study of the physical

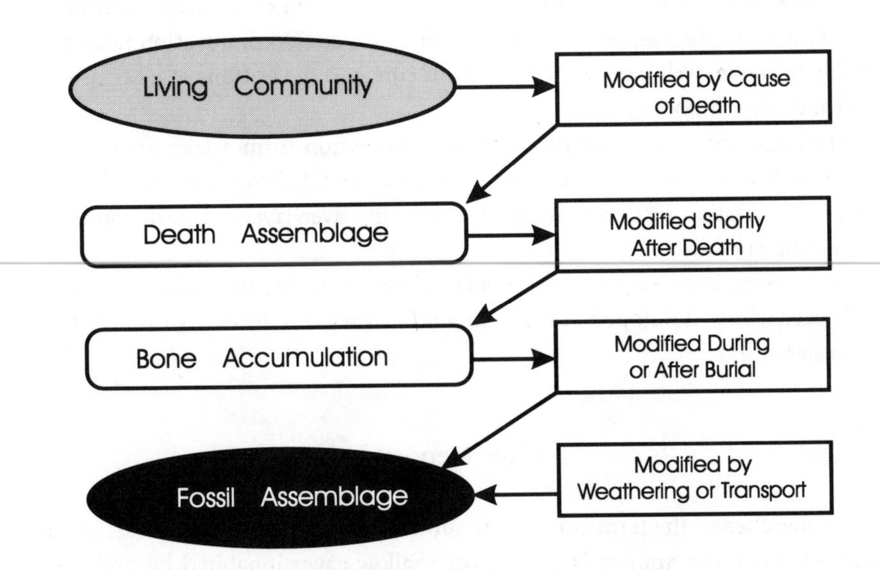

Figure 2.3. Summary of bone taphonomy processes. (Modified from Andrews, 1990.)

and chemical properties of the sediments themselves. The size of the particles in the sediment (gravel, sand, silt, and clay) reveals whether they were deposited in a low-energy environment (in the lake, or from a slowly moving stream) or a high-energy environment (from a river, a flooding creek, or a beach gravel redeposited after a severe storm). Most deposition takes place in low-energy environments, whereas erosion takes place in high-energy environments.

The chemical characteristics of lacustrine sediments dictate the types of fossils that will be preserved in them. **Alkaline sediments,** such as marls, preserve bones and mollusk shells far better than do **acidic sediments.** This is so because acids attack the calcium carbonate that forms the bulk of these materials.

Peat Bogs

The two principal types of peats are moss peats (frequently dominated by sphagnum mosses) and sedge peats. Moss peats accumulate in bogs, whereas sedge peats accumulate in fens. Peat bogs are fed by precipitation and are acidic; they are also poor in nutrients. The water in fens often comes from a combination of precipitation and ground water; they are often less acidic than bogs and richer in nutrients.

Peat forms when the remains of aquatic or semiaquatic plants in a bog or fen accumulate faster than they decompose. This process is helped along by the water in the bog or fen. Plant remains that sink to the bottom of the water decompose much more slowly than plant remains in the open air. Decomposition takes place readily in warm, well-oxygenated environments, but the bottom of a bog is often cold and oxygen-poor.

Bogs may form as a part of **ecological succession** from a lake or pond to a meadow. Many mountain meadows in Alaska are filled-in ponds or lakes. The modern vegetation of these meadows is growing over layers of peat, sometimes several meters thick.

Bogs cover large regions of Canada, Alaska, and Siberia today. During the Pleistocene, bogs developed in many unglaciated regions, leaving substantial peat deposits behind.

Cave Deposits

In a generic sense, the term *cave* covers subterranean caverns of all sizes; the term *rockshelter* is more appropriate for most shallow caves inhabited by prehistoric animals and people in Alaska and elsewhere. Rockshelters are shallow enough to allow considerable sunlight and air circulation, even at their deepest point, whereas caves may extend many miles underground and include regions of complete darkness and little exchange of air with the outside world. I use the term cave here in the more generic sense.

For our purposes, cave deposits are most important as a source of vertebrate fossils. Caves are essentially closed systems, limited to walls, floor, and roof. They have a beginning (cave formation) and an end (collapse of the cave roof or infilling by outside sediments). Such boundaries mean that the sediments (and associated fossils) that collect in a cave represent a discrete time interval. This process is in contrast to many other types of sediment deposition, such as those in large lakes or rivers, where sediments may accumulate for tens or hundreds of thousands of years. Accumulation of sediments in a cave begins when an opening forms to the outside world. Most caves that have produced abundant vertebrate bones are small, shallow caves that are close enough to the surface to have substantial contact with the world outside the subterranean cavern.

There are five sources of bones in cave deposits (Fig. 2.4):

1. Animals carried in by predators.
2. Fecal pellets of animals or birds frequenting or living in the cave.
3. Predators or scavengers that are attracted to carrion and then become trapped when they enter the cave.
4. Animals that fall into the cave through cracks or other openings from the surface.
5. Cave-dwelling animals.

Other than specialized cave dwellers, most animals will only live in caves that have easy access to the outside world. Pleistocene caves provided shelter for numerous animals, including humans. When these animals died in caves, their remains were often preserved in cave-floor sediments. However, cave researchers have concluded that most bones in cave deposits are from animals that accidentally fell in rather than from actual cave dwellers. A good example of this kind of cave is Natural Trap Cave in Wyoming; it has a small opening at the top, which widens to form oversteepened walls inside. Such caves often contain deposits with an abundance of carnivores (mountain lions, wolves, and other mammals). It is thought that the carnivores are attracted to the smell of carrion emanating from the opening at the top of the cave and then enter (or fall into) the cave themselves and are unable to get out.

Predators, including birds of prey, often carry prey animals into the open mouths of caves. The bones of the prey are then deposited in the cave either as refuse left over from the carnivore's meal, in regurgitated pellets, or in the predators' scat.

Figure 2.4. Generalized section of a cave, showing potential sources of fossil bones: 1, predators that carry prey animals into the cave; 2, owl and other raptor pellets; 3, predators attracted by carrion that fall into the cave through a large vent; 4, animals that fall into the cave through vents; 5, animals that live in the cave. (Modified from Andrews, 1990.)

Depletion of Fossil Resources

There are a wide variety of sources for Quaternary fossils. These fossils are all around us, literally underfoot. Unfortunately, many valuable sources of Quaternary fossils are being exploited by people for other purposes. For example, peat is mined from now-shrinking deposits in Canada, the northern United States, Europe, and the Soviet Union. Peat is used both as organic matter to improve garden soils and, in some places, as a fuel for heating and cooking (albeit not a very good one). Ancient fluvial deposits are mined as sources for sand and gravel. Although such mining activities have brought many Pleistocene fossils to light through the years, they have devoured many more. Ancient mammoth ivory has been used extensively for jewelry, buttons, and even piano keys and billiard balls. Artifacts from archaeological sites, such as arrow- and spearheads, pottery, and stone carvings, are taken away by "pot hunters." Yet these materials are the keys that unlock the book of prehistory; it is a shame when they are reduced to being just another traded commodity or collectable curiosity.

Suggested Reading

Andrews, P. 1990. *Owls, Caves, and Fossils.* Chicago: University of Chicago Press. 231 pp.

Birks, H. J. B., and Birks, H. H. 1980. *Quaternary Palaeoecology.* London: Edward Arnold Publishers, Ltd. 289 pp.

Guthrie, R. D. 1990. *Frozen Fauna of the Mammoth Steppe. The Story of Blue Babe.* Chicago: University of Chicago Press. 323 pp.

3

DATING PAST EVENTS

Once we have assembled bits and pieces of information from various types of fossils, the next step in reconstructing past environments is to fit the fossil data into a time frame or chronology. This procedure may be done in a number of different ways, depending on the types of materials available for dating and the interval of time we wish to study (Fig. 3.1). Accurate dating is essential to paleoecology; without it, it is impossible to determine the rates at which past environmental changes took place (for example, did a climatic warming begin rapidly or more slowly?). Accurate dating also makes it possible to determine whether past events took place at the same time across a broad region or whether those events were unrelated to each other.

In order to explain most Quaternary dating methods, we must make a brief excursion into the highly technical, often baffling realms of high-energy physics and organic chemistry. Although some readers may relish the opportunity to tackle **isotopes,** ions, and isomers, others may shy away from this chapter. For me, these techniques are simply a means to an end . . . getting reliable dates on fossil materials. However, it always helps to understand the principles behind the technology one relies on, even if that understanding would be considered rudimentary by experts in the field. I have attempted to provide an overview of the major methods, explaining just

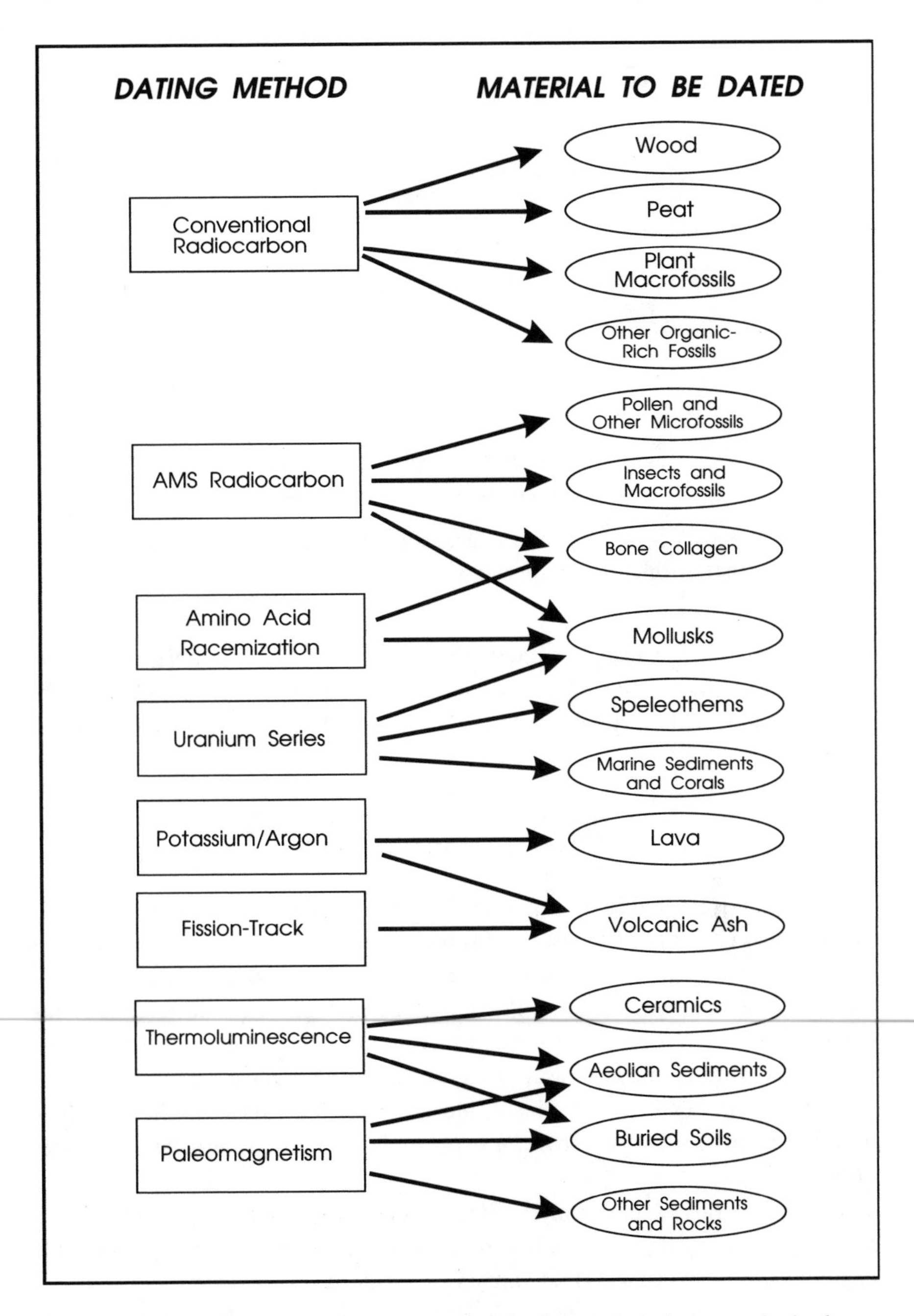

Figure 3.1. Summary of radiometric, chemical, and paleomagnetic dating methods, showing the types of fossils and other materials dated by the methods.

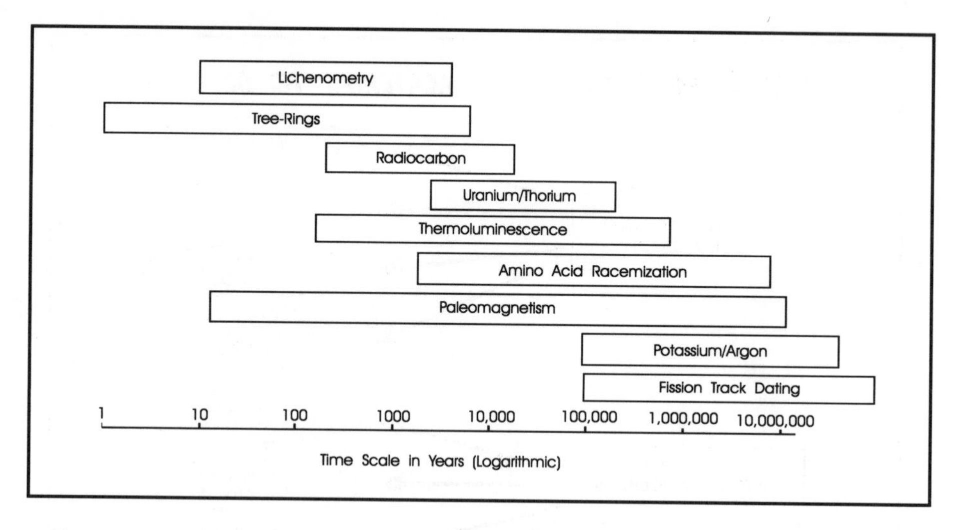

Figure 3.2. Summary of radiometric, chemical, and paleomagnetic dating methods, showing the time ranges each method is capable of measuring.

enough to allow the reader to grasp the idea without getting bogged down in endless details that dating specialists must deal with on a day-to-day basis. For additional information, please consult the suggested readings at the end of the chapter.

Types of Dating Methods

Dating methods fall into four categories: (1) radiometric methods; (2) paleomagnetic methods; (3) chemical methods; and (4) biological methods. Radiometric methods measure the radioactive decay of unstable isotopes of atoms, including isotopes of carbon, potassium, and uranium. Paleomagnetic methods measure changes in the magnetic field of the earth, as revealed in changes in polarity of magnetically charged atoms. Chemical methods measure time-dependent changes in certain chemicals, including chemicals in fossils. Finally, biological methods measure the growth of long-lived plants, especially trees (as expressed in tree rings) and certain species of lichens.

Each of the various dating methods is useful over a certain time span (Fig. 3.2). Some are used to date events only within the last few thousand years; others only begin to date events 10,000–15,000 years old or older.

The principal radiometric methods used in Quaternary studies are based on the decay of radiocarbon to stable carbon (^{14}C to ^{12}C), potassium/argon dating, and uranium series dating. I discuss each method in turn.

Radiocarbon Dating

Radiocarbon dating is the most widely used dating method for late Quaternary fossils. The ^{14}C isotope of carbon is continuously being created in the upper atmosphere by the bombardment of nitrogen atoms by neutrons from cosmic rays emanating from the sun (Fig. 3.3). The neutrons react with stable nitrogen atoms (^{14}N) to create a radiocarbon atom and a hydrogen atom. The radiocarbon atoms rapidly combine with oxygen to form $^{14}CO_2$, which diffuses down through the atmosphere and is taken up by plants in **photosynthesis.** The plants store the ^{14}C atoms in their tissues, along with much larger quantities of stable (^{12}C and ^{13}C)

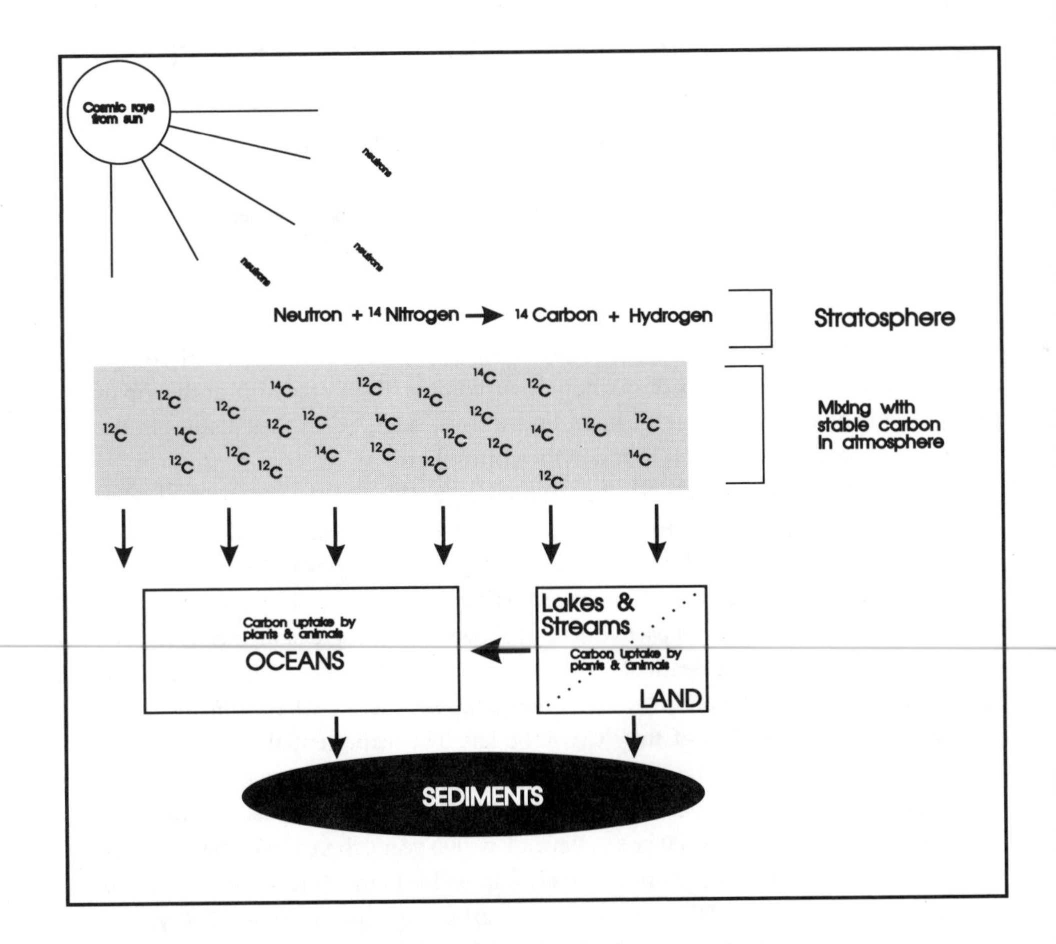

Figure 3.3. Diagram summarizing how radiocarbon is created and spreads in the atmosphere, its uptake by living organisms, and final deposition in sediments. The arrows show the direction of movement of radiocarbon in the system.

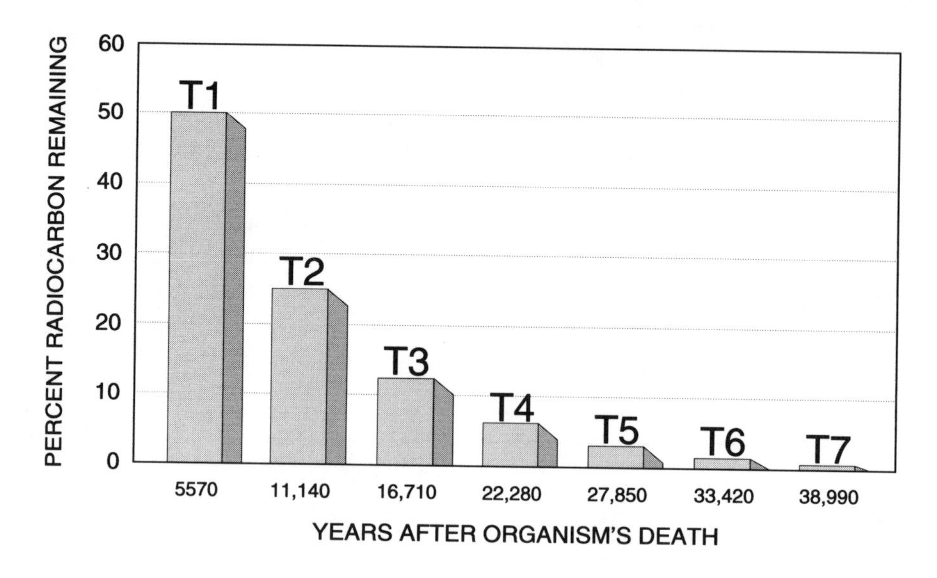

Figure 3.4. Bar graph showing the decay of radiocarbon through time, expressed in terms of the percentage of original radiocarbon atoms remaining in a sample through seven half-lives.

carbon. The ^{14}C atoms are continuously taken in by plants and, secondarily, by the animals that eat the plants throughout their lifetime. Even predators at the top of the food chain (such as eagles, lions, and wolves) have ^{14}C in their tissues in the same proportion to ^{12}C as is found in the atmosphere.

Once plants and animals die, they stop taking in ^{14}C, and the ^{14}C in their bodies begins to deplete. Because the ^{14}C atom is radioactive, it begins to decay back to nitrogen (^{14}N). The rate of this decay was determined by Libby in 1955. Half of the ^{14}C atoms will have decayed to nitrogen within 5570 years. This time interval is called the *half-life* of the radioactive decay. Following the progression, then (Fig. 3.4), by 11,140 years after the death of an organism, only 25% of the original ^{14}C content will remain. By 16,710 years after death, only 12.5% of the ^{14}C will remain. By 22,280 years after death, 6.25% of the ^{14}C will be left. This exponential decrease in the amount of remaining radiocarbon can be measured to approximately 10 half-lives, or 55,700 years before present (B.P.). In practice, ^{14}C dating is reliable only for samples younger than 40,000 yr B.P. Beyond 40,000 years, there is so little radiocarbon left in a fossil that it becomes virtually impossible to measure accurately. Fossils older than 40,000–45,000 years yield radiocarbon ages of 40,000–50,000 yr B.P, whether they are 50,000 years old or 50,000,000 years old. Radiocarbon laboratories designate such samples as, for instance, "greater than 40,000 yr B.P."; this is then taken as the minimum age of the sample, since it could be much older.

There are now two methods of obtaining radiocarbon ages from a sample. For large samples of organic material (twigs, logs, or blocks of peat), conventional methods may be used, including either gas-counting or liquid scintillation. For the gas-counting method, the sample is burned, and the gases emitted from burning (such as carbon dioxide and methane) are placed in a chamber with a detector that counts emissions of β **particles** (electrons given off in the radioactive decay process). The liquid scintillation technique involves converting the organic component of a sample into a liquid (benzene or some other organic liquid). This liquid is placed in a chamber that detects scintillations, the minute flashes of light given off when a β particle is emitted from the liquid.

Within the last decade, the development of the accelerator mass spectrometer (AMS) method has allowed much smaller samples to be radiocarbon dated. Whereas the conventional methods require samples containing at least several grams of carbon to yield reliable dates, the AMS method is able to date minuscule samples such as individual seeds, insect parts, and tiny lumps of charcoal. For instance, the AMS method was used recently to obtain a ^{14}C age on single strands of fabric from the Shroud of Turin, which was claimed by some to have been the burial shroud of Christ. In fact, the AMS dates showed that the Shroud of Turin was made during medieval times. Instead of measuring β particle emissions, which are an indirect measure of the amount of ^{14}C in a sample, the AMS method measures the actual concentrations of the carbon ions (^{14}C, ^{13}C, and ^{12}C) in a sample. The **ions** are accelerated in a **cyclotron** or tandem accelerator chamber to extremely high velocities. Then they are passed through a magnetic field, which separates the different ions, allowing them to be distinguished from each other and counted individually.

Potassium/Argon Dating

Potassium/argon dating is based on the radioactive decay of ^{40}K to ^{40}Ar. Potassium commonly occurs in two stable isotopes (^{39}K and ^{41}K) plus small amounts (0.012%) of the unstable isotope ^{40}K. In comparison with radiocarbon, ^{40}K decays very slowly. The half-life of the decay of ^{40}K to ^{40}Ar is 1.31 billion years. Argon is a gas. It can be driven out of a rock sample by heating. The K/Ar method is used to date volcanic rocks in which all initial argon was driven out when the lava was in a molten state. As time passes, ^{40}Ar builds up in the lava again from the decay of radioactive potassium. By measuring both the ^{40}K and ^{40}Ar content of a lava sample, the ratio of potassium to argon can be found, and the number of half-lives expended can be calculated. However, because the radioactive decay is so slow, this method is useful only for samples more than 100,000 years old.

Uranium Series Dating

Uranium series dating is based on the radioactive decay of uranium, either the ^{238}U or the ^{235}U isotope. The ultimate product of uranium decay is stable lead (^{206}Pb or ^{207}Pb). However, the process involves several intermediate isotopes of various elements, called daughter products of the radioactive decay. For instance, one of the principal daughter products of the decay of ^{238}U is the thorium isotope ^{230}Th, which has a half-life of 75,200 years. One of the daughter products of the decay of ^{235}U is the protactinium isotope ^{231}Pa, which has a half-life of 32,400 years. These two daughter products are the main isotopes studied in uranium series dating. One important feature of the thorium and protactinium isotopes is that they do not dissolve in water. This means they **precipitate** out of solution and collect in sediments. Because the concentration of uranium in sea water is a constant, its accumulation rate in sea sediments is also known, and the age of the sediment can be estimated, based on the percent of uranium that has decayed into the daughter products.

The useful dating range for $^{238}U/^{230}Th$ is 10,000–350,000 years. The useful dating range for $^{235}U/^{231}Pa$ is 5000–150,000 years. Uranium series dating is used mostly for marine fossils such as corals. It is also used to date terrestrial fossils or features containing carbonates, such as speleothems (stalactites and stalagmites) from caves. The mineral skeletons formed by marine corals include uranium from sea water but almost no ^{230}Th. Therefore, the age of corals can be estimated on the basis of the buildup of ^{230}Th from the radioactive decay of ^{238}U.

Thermoluminescence

Thermoluminescence (TL) is the light emitted from a mineral crystal when it is heated following the mineral's exposure to radiation. Electrons that are produced by radioactive decay (β particles) become trapped in the crystal matrix of minerals. When the mineral is heated, the electrons escape and give off light. The longer the mineral has been in the ground collecting free electrons, the more light it will give off when heated. Since heating discharges the electrons from the crystals, it sets the TL "clock" back to zero. This makes TL useful for dating archaeological artifacts such as ceramic pottery. The clay minerals used to make a pot will discharge their TL completely when they are fired in a kiln. Any subsequent measurable TL can then be used to date the time elapsed since the ceramics were made.

Thermoluminescence only works for sediments or artifacts that are buried in complete darkness. This is because sunlight also empties the minerals of their TL and resets their TL clock. Buried loess (windblown silt) deposits and buried soils have been shown to retain their TL "charge" until they are dug up to be sampled.

The useful age range of TL dates spans the interval of roughly the last million years; the accuracy of TL dating needs further refinements.

Chemical Methods

One of the principal chemical dating methods is amino acid dating. Amino acids are the building-block molecules of proteins. They are made by all living organisms. The structure of amino acid molecules is asymmetric or lopsided: the left-hand side of the molecule does not match the shape of the right-hand side. This asymmetric property creates an optical effect when polarized light is passed through the amino acid molecules. One configuration, or isomer, of the amino acid will rotate the plane of light to the left. This isomer is called the *levo* (Greek for "left") form. Nearly all amino acids in living organisms are in a *levo* configuration. The other isomer causes light to rotate to the right. This form is called the *dextro* (Greek for "right") form. After an organism dies, the amino acids in its body, which began as *levo,* or "L," isomers, slowly begin to change to *dextro,* or "D," isomers. This process is not unlike radioactive decay, except that it represents a chemical change that stops when the isomers reach an equilibrium point, whereas radioactive decay proceeds in one direction until the supply of unstable isotopes is exhausted.

The process by which amino acids change from one isomer to the other is called *racemization.* Amino acid racemization occurs at different rates in different organisms and is also greatly affected (as are most chemical reactions) by temperature changes. Still, amino acid racemization is a useful technique for dating one set of fossils relative to another set of fossils of the same species. Corrections can be made for fossils from different localities that have been in sediments of different temperatures.

The other principal chemical dating method is the chemical aspect of ash dating, or *tephrochronology.* The minerals that make up volcanic ash vary from one volcano to another and from one volcanic eruption to another. Because of this, each eruption produces volcanic ash, or *tephra,* with a unique chemical composition, or "signature." This chemical signature is not used to date the timing of the eruption (there are other methods for doing that), but once a specific volcanic ash has been dated, the ash itself becomes a very useful tool for correlating regional sediments. For instance, the Norse people (also known as Vikings) landed on Iceland about A.D. 900, the same year as an Icelandic volcano erupted (Icelandic volcanos are *very* active). The ash from that eruption is called the Landnám tephra (Landnám is the Old Norse word for landing). Wherever that tephra is found preserved in sediments on Iceland, it provides a convenient horizon that marks the timing of human arrival on the island.

Tephras are mostly dated in one of three ways. Late Pleistocene and Holocene ashes have been dated by radiocarbon dating of associated organic material, such as charred wood fragments or peat layers in bogs that are directly overlain by the ash. Ashes older than about 40,000 yr BP are dated by two other methods. If the mineral content of the ash contains sufficient potassium, it may be potassium-argon dated. Ashes that contain volcanic glass or minerals such as zircons may be dated by the fission-track method. Fission-track dating is based on the decay of uranium isotopes. As these decay, they emit radiation in the form of **alpha** (α) and **beta** (β) **particles.** The energy released in this process causes two nuclear fragments to be thrown out into the surrounding material. The resulting gouge marks or paths in the minerals or glass are called *fission tracks* (tracks caused by nuclear fission). These tracks are extremely short (on the order of a hundredth of a millimeter). After the volcanic glass or mineral has been polished and chemically etched to bring out the tracks, they can be counted under a microscope. Then the glass is heated, which causes the surface to smooth over, or anneal, as it melts. Once the ancient fission tracks are gone, the glass is then exposed to radiation under controlled conditions in the laboratory. The new radiation produces new fission tracks as the result of fission of ^{235}U. The number of newly created ^{235}U fission tracks is proportional to the uranium content, enabling the ^{238}U content of the glass sample to be calculated. So, when all is said and done, fission-track dating is just an indirect method of uranium series dating, allowing the researcher to date very small glass shards from volcanic eruptions.

Biological Methods

The biological dating methods are more "user friendly." That is, they are easier to understand and don't involve much in the way of high-powered physics or chemistry (as a biologist, I'll admit to a healthy share of bias in this statement). There are two principal biological dating methods. One is tree-ring counting, or *dendrochronology;* the other is the measurement of the growth of certain lichens, or *lichenometry.*

Dendrochronology is based on the fact that trees lay down annual rings as they grow. Each ring is made up of a broad, light-colored band that represents growth in the spring and summer and a narrow, dark-colored band that represents lessened activity in the fall and winter. Trees tend to grow broader rings during "good" years (i.e., years with adequate warmth, moisture, and nutrients) and narrower rings during "bad" years (years of drought, disease, and cold summers). The pattern of broad and narrow rings is repeated in trees growing throughout forest regions. This allows tree rings to be matched, or correlated, from tree to tree, which in turn allows tree-ring chronologies to be developed over greater lengths of time than the lives of individual trees. Of course, tree-ring researchers seek out very long-lived trees, such as bristlecone pines, from which to take cores for their studies. The cores are extracted from small holes drilled into the tree by a tool known as an increment corer. These holes do not usually harm the tree; they are quickly filled in with resin, which seals the wound. Bristlecones and some other conifers may live many thousands of years. By piecing together tree-ring chronologies from living and dead trees (trees whose lives overlapped at some point in the past), chronologies have been extended back throughout the Holocene, or the last 10,000 years. Tree-ring dating is so reliable and accurate that it has been used as a method of calibrating radiocarbon dating. This is done by chiseling out pieces of wood from individual rings of known age and ^{14}C dating the wood.

Lichenometry is based on the assumption that certain species of lichens grow at very slow, predictable rates. When these lichens are found growing on such objects as boulders in a glacial moraine, they can be used as an indirect way of dating glacial movements, by dating how long those boulders have been in place. The idea here is that when a rock gets caught up in glacial ice, all the lichens growing on it are scoured off or die as they are buried below the surface of the ice. Once the glacier stops moving or retreats, it leaves the boulder behind, and the clean surface of the boulder rapidly becomes colonized by new lichens.

Many species of lichens grow radially; that is, they grow through a series of expanding circles rather than stretching along a line. By measuring the diameter of radial-growth lichens, it is possible to estimate their age (Fig. 3.5). An aid to this procedure is to measure the size of lichens growing on an object of known age. One favorite source of this information is lichens growing on gravestones. This is particularly useful in Europe, where gravestones date back many centuries. One of the most commonly used lichens for lichenometry is the **crustose lichen,** *Rhizocarpon geographicum.* Lichens can live for several thousand years, especially in cold climates. One problem with lichenometry is that lichen growth tends to slow down with age. Another is that it is exceedingly difficult to identify lichens accurately (they are, after all, a combination of an alga and fungus, growing **symbiotically**).

Many tools are used to obtain ages of sediments and fossils. Some are more precise than others, and all are applicable to only certain segments of the Quaternary time line. Often, they are used in combination. For instance, a radiocarbon age may be obtained from organic materials in silt, and a TL age obtained from the silt itself. However, none of these dating methods is cheap. The current charge for a conventional ^{14}C date is about $250, and an AMS radiocarbon date can cost more than $1000, depending on how much work is required to prepare the sample for dating; not many Quaternary researchers can afford to obtain more than a few dates from a given site. A common practice for samples rich in organic matter that

Figure 3.5. *Alectoria* lichen growing on a boulder in a glacial moraine, Baffin Island. (Photograph by Dr. Gifford H. Miller, reprinted by permission of Dr. Miller.)

are thought to be less than 45,000 years old is to obtain a radiocarbon date from a basal sample, one from near the middle of the stratigraphic section, and one from the top. When possible, researchers try to get additional dates on horizons representing times of environmental change, so that the timing of those changes can be pinned down more precisely.

Suggested Reading

Berglund, B. E. (ed.). 1986. Dating methods. Chapters 14–19 in *Handbook of Holocene Palaeoecology and Palaeohydrology.* New York: John Wiley & Sons. 869 pp.

Bradley, R. S. 1985. Dating Methods I and II. Chapters 3 and 4 in *Quaternary Paleoclimatology. Methods of Paleoclimatic Reconstruction.* Boston: Allen and Unwin. 472 pp.

Fritts, H. C. 1976. *Tree Rings and Climate.* New York: Academic Press. 567 pp.

Schweingruber, F. H. 1988. *Tree Rings: Basics and Applications of Dendrochronology.* Boston: Reidel. 276 pp.

4

PUTTING IT ALL TOGETHER

We have covered a lot of ground in the last three chapters. I have tried to bring the reader up to date on types of Quaternary fossils, where they are found, and how deposits are dated. In this chapter, I provide a summary of how these data are combined to reconstruct past environments. The job of synthesizing the data into a meaningful reconstruction of past environments is often the most difficult aspect of Quaternary science, but it can also be the most satisfying. After months of sieving samples, peering down a microscope, or boiling sediments in chemical baths for a pollen preparation, it is refreshing to step back from the mundane tasks and discuss the "big picture" with colleagues. The process may take weeks or months (sometimes even years) to complete, and the initial attempts to fit the data together may end in the painful decision to go out and get more or better samples before any of the important questions can be answered. The process takes patience, perseverance, and a broad outlook.

The first level of paleoclimatic reconstruction is the process of planning the research project. This originates by a determination of which research questions are important and answerable (or, at least, which hypotheses can be tested). Once the research plan is drawn up (and permission must be obtained from the National Park authorities if the work is to be done in a National Park), the scientists can begin collecting data. Data collection involves fieldwork to collect samples, followed by laboratory analyses of samples. The second level of research involves converting the raw data from the laboratory into paleoclimatic data. This process ranges from simple, qualitative approaches to complex, quantitative approaches. The simplest approach is to analyze the data through such means as finding the modern distribution of each species found in a fossil assemblage and then determining the geographic region where the modern distributions of the species in question overlap. Then the modern climate of that region is used as an estimate of what conditions were at the time and place associated with the fossil assemblage. Most sophisticated approaches involve statistical transformations of the fossil data, using mathematical formulas based on modern observations to fine-tune estimates of paleoclimates.

The third level of paleoclimate reconstruction involves combining a series of local studies into a regional synthesis. This type of study attempts to describe regional climatic patterns over a given time interval and then proceeds to compare the patterns derived from the proxy data with theoretical climate models, such as models that reconstruct general circulation patterns in the atmosphere. Synthesizing data is best done by a team of researchers, each of whom approaches the research questions from a different angle or discipline (e.g., climatology, paleontology, or ecology). This type of interdisciplinary research (work that crosses the boundaries between the scientific disciplines) is the most valuable, for it produces the most coherent answers, tested from many different perspectives. The nature of interdisciplinary research is that the scientists cooperate to arrive at an answer to one question by using different means, often from different disciplines.

Imagination is one aspect of paleoecological research that, surprisingly, is quite essential. You may be able to assemble all the facts and figures from a fossil assemblage, but unless you can recreate a prehistoric scene in your imagination, you probably will not put the data together in a very meaningful way. This does not mean that our analyses are just a bunch of daydreams; far from it! Rather, it means that researchers have to combine their knowledge of how things currently work in the living world with the assembled body of fossil data in order to develop more than a superficial understanding of past events. This process involves the principle of *uniformitarianism*, formulated by British geologist Charles Lyell in 1830. This principle is summed up by the saying "the present is the key to the past." It might equally well be said that the past and present are interlocking parts of the whole: they are inseparable keys to each other's understanding. All of our modern animals and plants are just the latest generation of species that began in the distant past. If we are to understand them well enough to preserve today's populations, we need to study their history . . . their ancestral lineages that trace back hundreds of thousands of years.

The remainder of this chapter provides an overview of how the various bits and pieces of ancient biological and physical data are combined, or synthesized, to form major paleoenvironmental reconstructions, sometimes called "the big picture."

Reconstructing Physical Environments

The history of the physical environment is an integral part of paleoecology, because plants and animals operate in the physical world and respond as much to changes in the physical environment as they do to biotic interactions (e.g., competition for resources, predation, parasitism). The physical environment can be broken down into three parts: the land (the geosphere), the water (the hydrosphere), and the air (the atmosphere). Obviously, the three elements interact continuously with each other. Nearly all of the energy that drives these interactions ultimately comes from the sun.

Changes in the physical environment are recorded in the features of ancient landforms. In cold regions, such features as **glacial moraines** and ancient permafrost features are evidence of past glacial and **periglacial** environments. Permanently frozen ground leaves several types of evidence on a landscape, including **ice wedges, patterned ground,** and **stone stripes.**

Ice wedges form when then ground contracts in cold temperatures (–15 to –20°C, or 5 to –4°F), causing cracks at the surface. The cracks are expanded when water fills them and freezes. As long as temperatures remain cold enough, the ice wedge continues to widen at the top and deepen into a V shape.

Patterned ground is another feature of permafrost landscapes. There are two types of patterned ground. One is caused by the melting of ground ice. This type occurs at high latitudes, where permafrost has persisted throughout most of the Quaternary. Troughs in the ground form as the tops of ice wedges melt out during summer. This effect gains momentum in regions with a great deal of standing water, which contributes to the melting of ground ice. The troughs form a roughly circular net or polygon; the diameter of the polygon can be several meters, and the center may be raised into a mound or depressed into a hollow (Fig. 4.1). When this phenomenon spreads across a landscape, the net result is a large series of polygons that give some arctic regions a honeycomb appearance. Cracks that form in patterned ground may also become ice wedges through repeated widening and deepening by the freezing and thawing of ice.

The other type of patterned ground is created by frost action in the soil. This occurs in mountainous regions, even far south of the continuous permafrost zone. **Frost-heaving** over long periods of time tends to sort the material on or near the surface,

Figure 4.1. Ice wedge polygons forming patterned ground on the North Slope of Alaska. (Photograph by Dr. Donald Walker, INSTAAR, reprinted by permission of Dr. Walker.)

pushing cobbles and pebbles into ridges that form a polygonal net or series of rings. On steep hillsides, gravity forces the shape of the sorted material into a series of stripes.

Information on past glaciations can be derived through the study of shifts in the elevation of snowlines and glaciation thresholds. The boundary between regions where winter snows melt off a glacier in summer and regions of permanent snow is called the *equilibrium line altitude*, or the ELA. Above this line, snow continues to accumulate. It compresses under its own weight and forms ice, adding to the mass of the glacier. Below that line, snow does not accumulate year after year, and ice does not build up.

ELAs tend to fall on a climatic boundary that corresponds to regions in which summer temperatures average below the freezing point, 0°C (32°F). However, different climatic conditions seem to control ELAs in different regions. Obviously, precipitation plays a major role. No matter how cold it is, if there is no new snow, new ice will not form. During glacial intervals, temperatures were depressed to the extent that snow accumulated not just on mountaintops but across much of the high-latitude regions, eventually forming ice sheets that covered most of the northern regions and spread south to the midlatitudes. In mountains, glaciers advanced when the amount of ice buildup above the ELA was greater than the amount of ice melt-out below the ELA.

Past movements of glaciers are difficult to trace. Glaciers transport debris to their margins as they move. Where the glaciers from two mountain valleys come together, they frequently leave a pile of debris along their junction as well. When the ice retreats, moraines are left behind as testimony to past ice advances. Moraines consist of mounds of glacial debris: unsorted silt, sand, pebbles, cobbles, and boulders. They provide obvious evidence of glaciation.

There are two major problems, however, with trying to reconstruct past glacial events from morainal evidence. One is that each new glacial advance tends to obliterate the evidence from previous events. So, the glacier that advances the farthest down a valley will grind up and redeposit the moraines of all previous shorter advances. The second major problem is in trying to obtain an accurate date for past glacial events. Two methods are commonly used. One is to obtain a ^{14}C age from soils that develop on moraines. This age will be a minimum age only, because it may take several centuries to develop a good organic soil on a moraine. Moraines have also been dated by obtaining ^{14}C ages on logs incorporated in their sediments. The second method is to date the time when the rocks in the moraine stopped moving by using lichenometry. As we have already seen, this technique is not very accurate, although it works better on more recent moraines (less than a few thousand years) than on older ones.

Other methods of reconstructing past physical environments include studies of past lake levels and the study of physical properties of lake sediments. Past lake levels can be deduced from ancient beach ridges or terraces that indicate the altitude of past shorelines. The quantity and type of sediments found in lake sediments can reveal information on changes in the lake itself and on changes in local environments, such as the timing of episodes of soil erosion, input of sediments from nearby glaciers, and other features.

Reconstructing Climate Change

Fluctuations in the amount of insolation (*in*coming *sol*ar radi*ation*) are the most likely cause of large-scale changes in Earth's climate during the Quaternary. In other words, variations in the intensity and timing of heat from the sun are the most likely cause of the glacial/interglacial cycles. This solar variable was neatly described by the Serbian cosmologist and mathematician Milutin Milankovitch in 1938. There are three major components of the Earth's orbit about the sun that contribute to changes in our climate (Fig. 4.2). First, the Earth's spin on its axis is wobbly, much like a spinning top that starts to wobble after it slows down. This wobble amounts to a variation of up to 23.5° to either side of the axis (Fig. 4.2A). The amount of tilt in the Earth's rotation affects the amount of sunlight striking

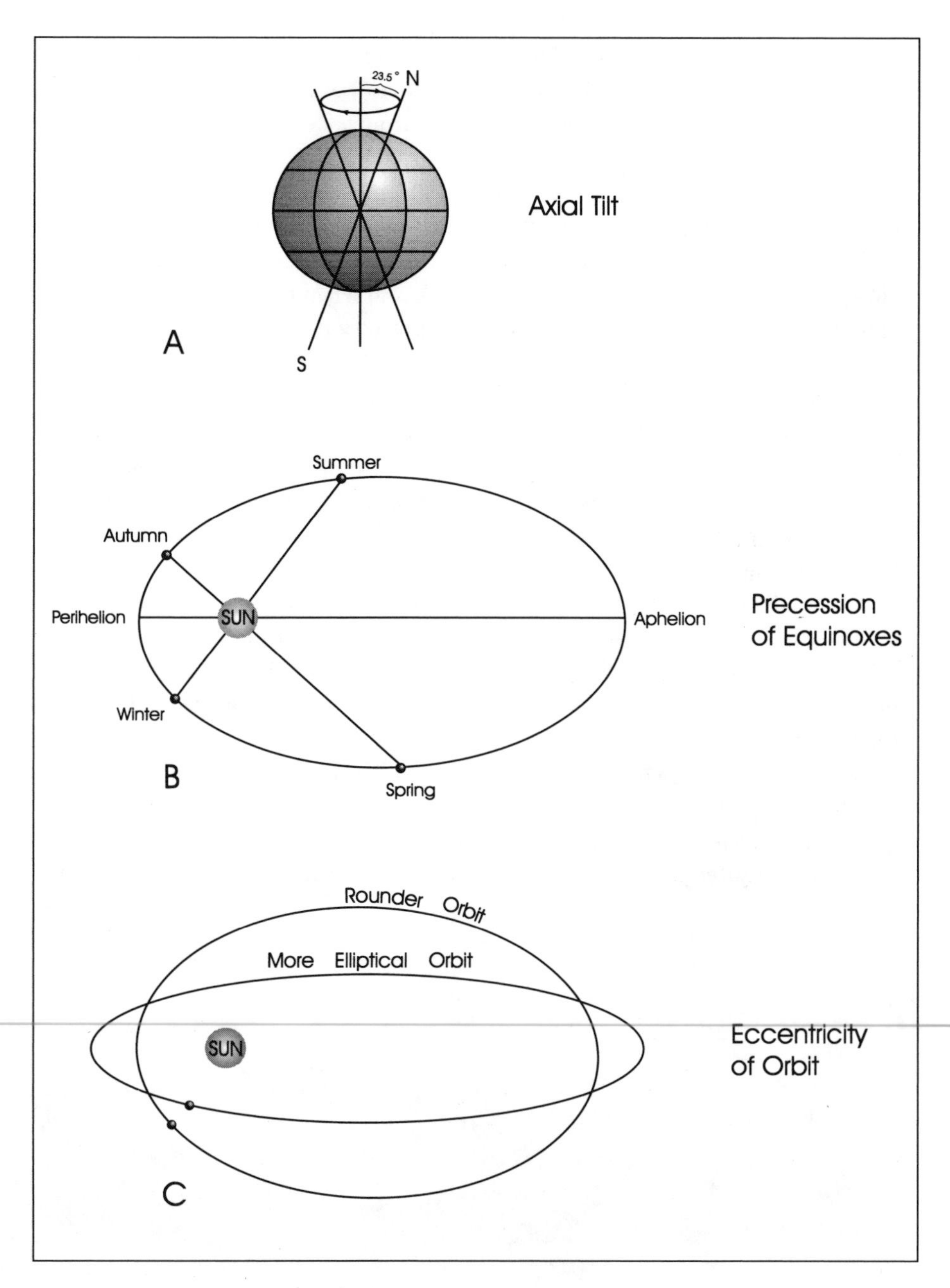

Figure 4.2. Illustration of the three major components of the Earth's orbit about the sun that contribute to changes in our climate, as described in the Milankovitch theory of ice ages.

the different parts of the globe. The greater the tilt, the stronger the difference in seasons (i.e., more tilt equals sharper differences between summer and winter temperatures). The range of motion in the tilt (from left of center to right of center and back again) takes place over a period of 41,000 years.

As a result of a wobble in the Earth's spin, the position of the Earth on its elliptical path changes relative to the time of year. For instance, in Figure 4.2B, autumn and winter occur in the northern hemisphere when the Earth is relatively close to the sun, whereas summer and spring occur when the Earth is relatively far from the sun. This phenomenon is called the precession of equinoxes. The cycle of equinox precession takes 23,000 years to complete. In the growth of continental ice sheets, summer temperatures are probably more important than winter. Throughout the Quaternary period, high-latitude winters have been cold enough to allow snow to accumulate. It is when the summers are cold (i.e., summers that occur when the sun is at its farthest point in Earth's orbit) that the snows of previous winters do not melt completely. When this process continues for centuries, ice sheets begin to form.

Finally, the shape of Earth's orbit also changes. At one extreme, the orbit is more circular, so that each season receives about the same amount of insolation. At the other extreme, the orbital ellipse is stretched longer, exaggerating the differences between seasons (Fig. 4.2C). The eccentricity of Earth's orbit also proceeds through a long cycle, which takes 100,000 years.

Major glacial events in the Quaternary have occurred when the phases of axial tilt, precession of equinoxes, and eccentricity of orbit are all lined up to give the northern hemisphere the least amount of summer insolation (Fig. 4.3). Conversely, major **interglacial** periods have occurred when the three factors line up to

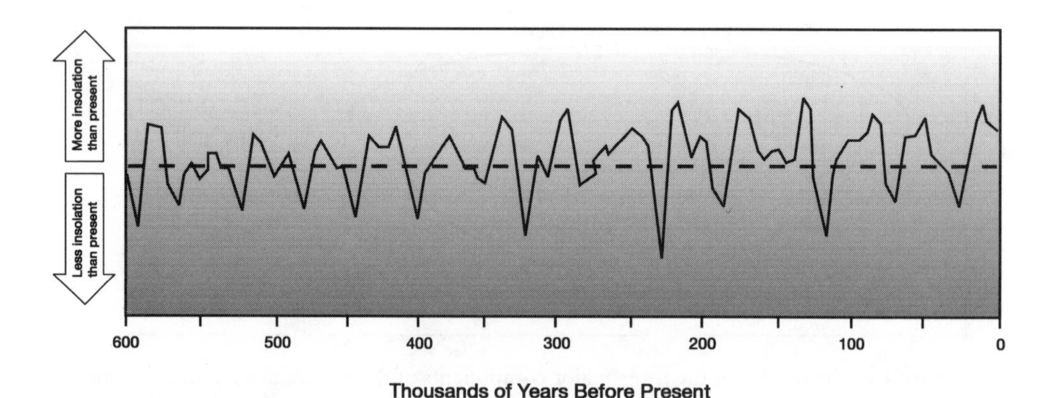

Figure 4.3. Insolation curve for the last 600,000 years at the latitude 65° N (about the latitude of Fairbanks, Alaska), based on the Milankovitch model.

give the northern hemisphere the greatest amount of summer insolation. The last major convergence of factors giving us maximum summer warmth occurred between 11,000 and 9000 years ago, at the transition between the last glaciation and the current interglacial, the Holocene.

Paleoclimatic reconstructions involve piecing together data from fossils and other physical evidence and combining them with a chronology based on stratigraphy and various types of dating. As with all theoretical science, there are inherent problems in paleoclimatic reconstruction. One is that each type of proxy data responds to climate in its own way and at its own rate. For instance, some types of plants (especially trees) may take centuries or millennia to respond fully to a major climatic change, whereas insects may respond to the same change within a few years or decades. In this scenario, pollen and insect fossils from the same lake sediment samples tell very different stories. Such was the case for samples from the end of the last glaciation in Britain. The fossil insect faunas showed that summer temperatures were very close to modern levels, but the pollen records were still registering arctic tundra plants. Eventually, the conifers and then the hardwood trees arrived, but not until many centuries later.

Another problem in paleoclimatic reconstructions based on proxy data is that some data are more or less continuously deposited, whereas other data are discontinuous or spotty. For instance, a lake basin may collect sediments every year for thousands of years, whereas an adjacent glacier leaves moraines that date only to one century.

Reconstructing Ecosystems

Reconstructing ancient climates may seem difficult, but it is relatively straightforward compared to reconstructing past ecosystems. The reason is simple: plants and animals behave in much more complicated and unpredictable ways than sediments, glaciers, and frozen ground. If individual species responded independently to their environment, paleoecology would be simpler. Unfortunately, the species are constantly interacting with each other in ways that are very hard to see in a fossil record.

Like paleoclimate studies, paleoecological studies can be divided into first-level, or descriptive approaches, and second-level, or statistical approaches. The second-level approach is gaining in popularity and success. It employs numerical and computer techniques and can sometimes reveal subtle patterns in the data that would otherwise be missed. Statistical methods can also be used to standardize the paleoecological reconstructions of several researchers working in a given region.

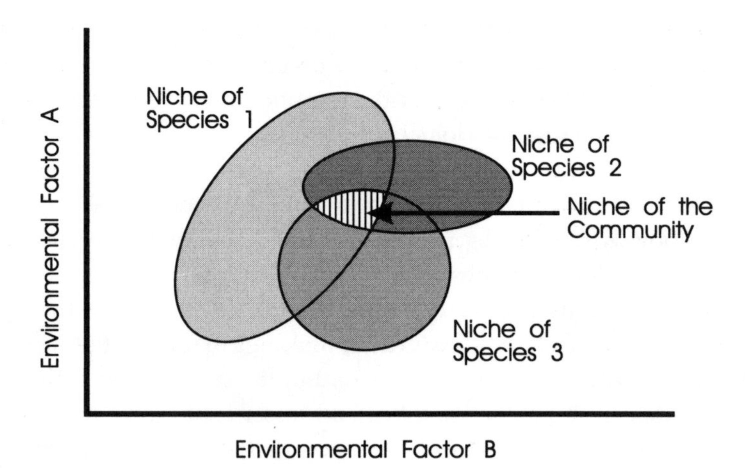

Figure 4.4. The niche of a biological community, illustrated in a hypothetical community of three species and their ecological requirements for two environmental factors. (After Birks and Birks, 1980.)

In order to reconstruct past ecosystems from fossil data, paleoecologists must develop studies that sample well-preserved fossils in the appropriate biological groups from the right time interval to answer the research question. Most paleoecologists are trained to identify fossils in one or a few groups of plants or animals. There are very few generalists in paleoecology, because virtually no one has the time and energy to learn to identify all the types of fossils found in a deposit. For instance, in order to identify fossil insects from sites in North America, it is necessary to develop a good working knowledge of about 100 different families of beetles, ants, and other groups.

Once a group of fossils has been identified from an assemblage, it is necessary to explore the ecological requirements of their modern counterparts (assuming the species in the fossil assemblage have not become extinct). Each species is adapted to a range of environmental tolerances and forms a part of a biological community. The factors that control its position within that community are described as its *niche.* A gathering of data on individual species' ecological requirements provides a good deal of paleoecological information, but a clearer picture can be had by reconstructing whole communities, because the community itself occupies a certain niche within a geographic region, and that niche is more narrowly defined than the niches of individual species (Fig. 4.4). For example, an alpine tundra community may be able to exist only on south-facing slopes of well-drained hills in central Alaska, even though several species that are a part of that community are able to live in a variety of habitats away from such slopes.

This principle is good in theory, but fossil records hardly ever preserve whole communities intact (for example, see Fig. 2.2). However, if enough elements in a community are found in a fossil assemblage, comparisons with modern communities may be possible. This process is also strengthened by the presence of *indicator species*, that is, species that are strongly indicative of certain communities. For instance, mastodons appear to have lived only in close proximity to coniferous forests (conifer needles were a major part of their diet). Wherever mastodon fossils are found, it is reasonable to assume that coniferous forests and an associated boreal climate were also present.

Finally, the reconstruction of past ecosystems is based on the compilation of fossil assemblages from a number of regional sites representing various communities. Ancient ecosystems are described with the aid of modern ecosystems that are analogous to the ancient ecosystems. Comparing and contrasting modern analogues with ancient biotas fill in some of the gaps while showing differences in the structure and function of past versus present ecosystems. No ancient ecosystem was exactly like any modern ecosystem. This is because ecosystems are constrained by the conditions of the physical environment, which has changed continuously through time. The other disparity is that past communities were made up of unique mixtures of species. Even though there can be similarities between communities through time, species migrate, become established in new regions, and die out in others. All of this is in response to changing environments, competition between and among species, and changing resource availability.

Archaeology: How People Fit the Picture

For the most part, humans had less impact on North American ecosystems than they did on ecosystems in Europe and Asia, at least until the Europeans arrived on the scene beginning in 1492. This is helpful to paleoecologists working on North American sites. In contrast, our European counterparts are often faced with the dilemma of trying to separate human modifications of past landscapes from natural changes brought about by climate change or other factors. In other words, anthropogenic (human-induced) effects on European landscapes date back several thousand years, confounding attempts to reconstruct natural environments and ecosystems.

One of the biggest remaining mysteries in the study of North American ecosystems concerns the mass extinction of large mammals, or megafauna, at the end of the last ice age, 11,000 years ago. One school of thought holds that early hunters (Paleoindians) wiped out most of the North American megafauna, shortly after they arrived from Asia via the Bering Land Bridge. Another theory holds that the megafauna became extinct because their habitats were greatly disrupted as environments changed at the end of the ice age. Some scientists believe that the megafauna was on its way out because of environmental factors and that the Paleoindians merely delivered the *coup de grâce.* We will discuss this issue more fully in Chapter 6.

The fields of archaeology and paleoecology collaborate under the banner of *geoarchaeology.* Research in this field attempts to fit archaeology more closely into a paleoenvironmental scenario. Fossil data are collected from the archaeological site or from adjacent natural deposits or both. Moreover, geoarchaeologists use fossil data (rather than just archaeological artifacts) to help develop an understanding of prehistoric peoples and their ways.

Suggested Reading

Birks, H. J. B., and Birks, H. H. 1980. *Quaternary Palaeoecology.* London: Edward Arnold Publishers, Ltd. 289 pp.

Bradley, R. S. 1985. Chapter 7: Non-marine geological evidence. In *Quaternary Paleoclimatology. Methods of Paleoclimatic Reconstruction.* Boston: Allen and Unwin. 472 pp.

Dearing, J. A., and Foster, I. D. L. 1986. Lake sediments and palaeohydrological studies. In Berglund, B. E. (ed.), *Handbook of Holocene Palaeoecology and Palaeohydrology.* New York: Academic Press, pp. 67–90.

Digerfeldt, G. 1986. Studies on past lake-level fluctuations. In Berglund, B. E. (ed.), *Handbook of Holocene Palaeoecology and Palaeohydrology.* New York: Academic Press, pp. 127–131.

Dixon, E. J. 1993. *Quest for the Origins of the First Americans.* Albuquerque: University of New Mexico Press. 154 pp.

Holliday, V. T. (ed.). 1992. *Soils in Archaeology: Landscape Evolution and Human Occupation.* Washington, D.C.: Smithsonian Institution Press. 254 pp.

Rapp, G., Jr., and Gifford, J. A. (eds.). 1985. *Archaeological Geology.* New Haven: Yale University Press. 435 pp.

PART TWO

The National Parks of Alaska

ALASKA BOASTS MORE wilderness than any other state. Indeed, there is more wilderness in Alaska than in many of the other 49 states combined. The land set aside for National Parks in Alaska is bigger than entire states in the eastern United States. That alone would set the Alaskan National Parks apart from those of the "lower 48," but the Alaskan parks are wondrous places, even by Alaskan standards. They represent some truly fascinating ecosystems, and part of the fascination of these ecosystems is their unique history.

Much of Alaska today looks as it has for several thousand years, and probably as it looked during previous interglacial periods. A great deal of the land remains unspoiled and untamed. Because of its geographic isolation and forbidding winter climate, Alaska is sparsely populated. The entire human population of Alaska would fit comfortably in a single suburb of a major American city. Most Alaskans like it that way. Much of the state is unsuitable for agriculture and ranching, and the black spruces that blanket much of the interior are too scrawny to attract the attention of commercial loggers. These factors combine to keep most of Alaska unblemished by human activity. Where else could you go just a few miles outside the biggest cities and see wolves, grizzly bears, and bald eagles?

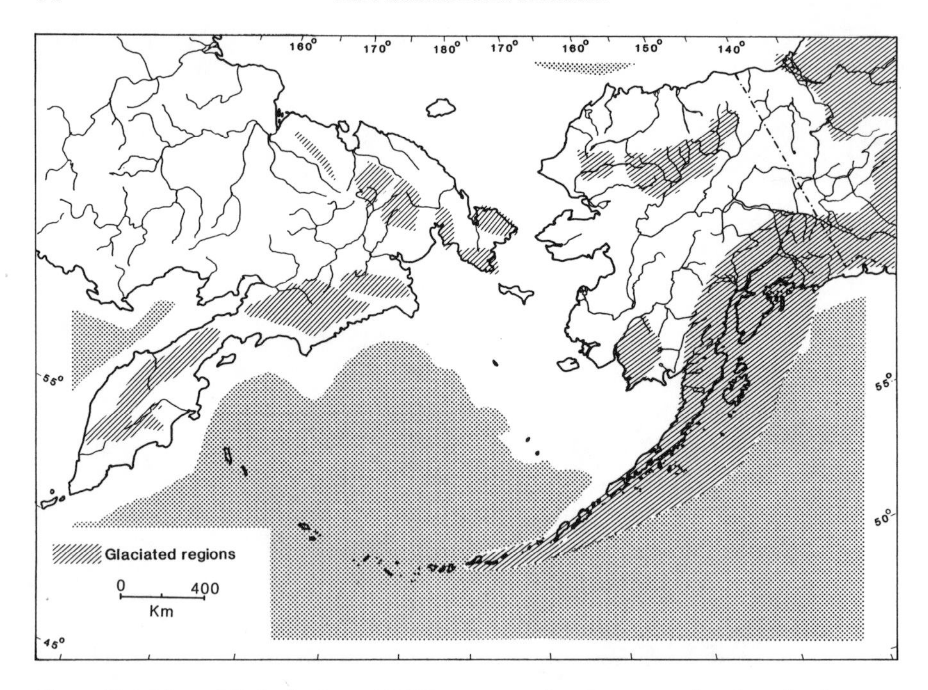

Figure II.1. Map of Beringia about 18,000 years ago, showing exposed regions, land bridge, and glaciers. (After Barry, 1982.)

This is not to say that people have not made any impacts on the land. Even the Denali region was bustling with miners and trophy hunters before the National Park was established. But the land is so huge, it dwarfs human impact. Archaeological research shows that people have been in Alaska for about 12,000 years, but they have left few lasting marks on the landscape until this century (or perhaps we just do not yet know how to see their impacts). It is hard for people to get a strong foothold in a **muskeg** bog or to dig too deeply into permanently frozen ground.

Alaska, the Yukon Territory of Canada, and eastern Siberia share a unique ice-age history. Even though these regions encompassed major arctic and subarctic terrain, their lowlands remained unglaciated during much of the Pleistocene. Sea level was lower during the glaciations, when ice covered most of the high-latitude regions (the water that would otherwise fill the seas was locked up in continental ice sheets). The continental shelves between Alaska and Siberia are very shallow; they are mostly less than 100 m below modern sea level. So, during the Pleistocene when sea level was lower, those shelves were exposed as dry land for tens of thousands of years. The combined unglaciated lands of Siberia, Alaska, the Yukon, and the intervening continental shelves formed a region known as *Beringia* (Fig. II.1).

Beringia was the major refuge (a refugium, in biologist's terminology) for arctic wildlife during much of the Pleistocene. Some populations of arctic plants and animals migrated south along the margins of ice sheets. They ended up in such exotic places as Iowa and Ohio. However, when the climate warmed, and the ice sheets began to melt, those populations died out in the south. Some arctic species were able to migrate back to the north with the receding ice. Those that could helped to recolonize the barren ground of northern Canada that was freed of ice during the Holocene. However, much of the recolonization of Canada came from populations of plants and animals that survived the last glaciation in Alaska and the Yukon.

As might be expected, the environmental conditions that developed in Beringia during Pleistocene glaciations were unique and unlike any that we can observe today. This is because the environmental factors that contributed to the formation of Beringia (lowered sea level and continental ice sheets) are no longer in existence—at least not until the next ice age comes along. Accordingly, the fossil record of Pleistocene Beringia tells of ecosystems we cannot visit.

Paleoclimate researchers have tried to single out the factors that played the most important roles in controlling the ice-age climatic conditions in Eastern Beringia (Alaska and the Yukon Territory). On the basis of computerized climate models, the important factors for this region, summarized in Figure II.2, included the

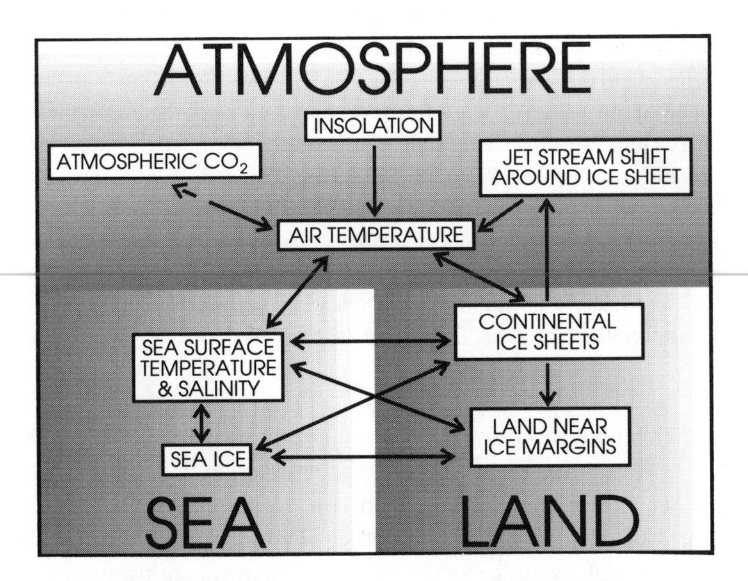

Figure II.2. Summary of climatic interactions among atmosphere, sea, and land that influenced the climate of Beringia during the last glaciation. (Based on Bartlein et al., 1991.)

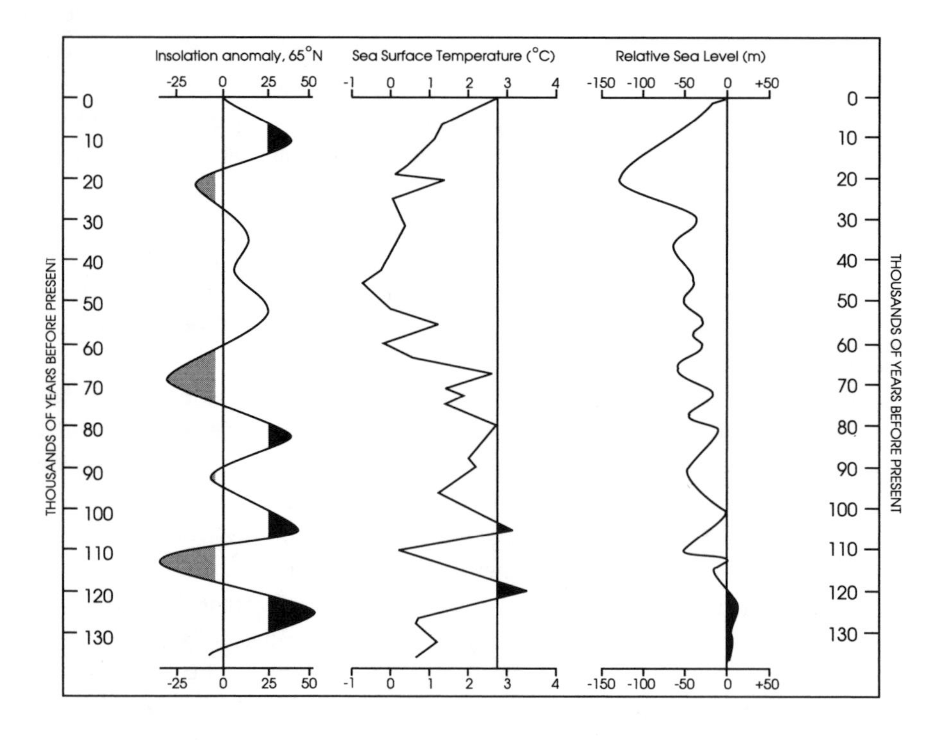

Figure II.3. Summary diagram showing predicted changes in insolation at high latitudes in the Northern Hemisphere, estimated changes in sea surface temperature for the north Pacific, and estimates in relative sea level for the Beringian region since the last interglaciation. Gray shading to the left of the vertical axis on the insolation anomaly graph indicates intervals when summer insolation was less than at present. Black shading to the right of the vertical axis indicates intervals when insolation was greater than the mid-Holocene maximum, which occurred at 6000 yr B.P. Sea surface temperature estimates are for winter conditions, inferred from fossil data. The vertical axis between 2° and 3°C indicates modern winter sea surface temperatures. Black shading indicates intervals when sea surface temperatures were warmer than at present. Black shading to the right of 0 (modern sea level) on the sea level curve indicates intervals when sea level was higher than at present. (Modified from Bartlein et al., *Quaternary International* 10–12:73–83, 1991.)

following: (1) cooling effects of the continental ice sheets to the east; (2) changes in the flow of the jet stream in the atmosphere that were brought about by the ice sheets (the jet stream apparently split in two on a regular basis, with one part flowing north of the ice sheets and the other flowing south of them); (3) changes in insolation at high latitudes in the Northern Hemisphere through the various Milankovitch cycles; (4) changes in the concentration of carbon dioxide in the atmosphere (glacial intervals experienced a reduction in CO_2, whereas interglacial intervals experienced increased CO_2); (5) sea ice, sea surface temperature, and salinity changes; and (6) cooling effects of late-lying glacial ice on adjacent land areas. None of these factors occurred in isolation; rather, they interacted in myriad ways, only a few of which we can claim to understand.

Some of these factors had unique impacts on Beringia. For instance, the spilt flow of the jet stream probably brought a stronger flow of southerly air into Alaska and the Yukon Territory, which would have made summers cooler than otherwise but also may have made the winters warmer. The combination of sea ice, sea surface temperatures, and salinity of ocean waters off the southern coast of Beringia may have brought warmer autumns and winters than exist under the modern interactions of seas, ice, and land.

The effects of CO_2 as a greenhouse gas in the atmosphere have been widely publicized in the late twentieth century, as the burning of fossil fuels boosts the atmospheric concentration of this and other gases. Increased CO_2 traps more solar energy in the atmosphere, preventing it from reradiating back into space. What is more mysterious is the fact that CO_2 concentrations have also fluctuated dramati-

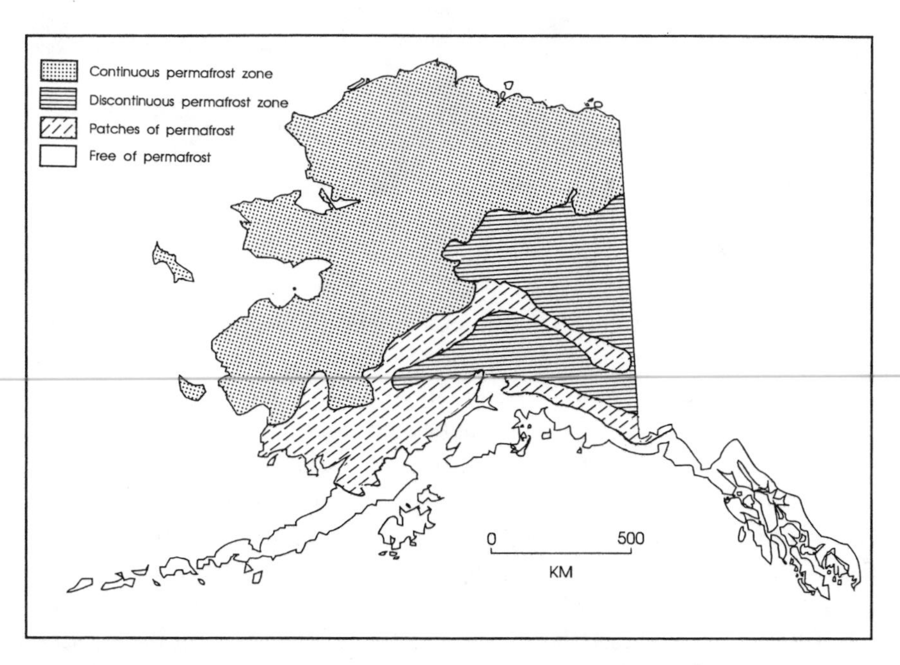

Figure II.4. Map showing the permafrost zones in Alaska. (Simplified from Johnson and Hartman, *Environmental Atlas of Alaska,* 1969, with permission of the University of Alaska, Fairbanks.)

Figure II.5. Photograph of an ice wedge, exposed near the mining town of Livengood, central Alaska. (Photograph by Oscar Ferrians, U.S. Geological Survey, used with permission.)

cally in the ancient past. No one knows exactly why this happened. Indeed, no one knew that it *had* happened until scientific technology became sufficiently advanced to allow the chemical analysis of tiny air bubbles trapped in ancient ice and other hard-to-study places such as air bubbles in amber.

The changes in three of these factors through the last glacial–interglacial cycle are shown in Figure II.3. This figure reveals several interesting patterns and linkages. For instance, the times of interglacial and **interstadial** warming correspond closely with peaks in insolation at high latitudes, which occurred at 130,000–120,000, 110,000–100,000, 87,000–75,000, and 60,000–30,000 yr B.P. and the Holocene (the last 10,000 years). At the other extreme, low points in the insolation curve correspond with the times of mountain glacial advances in Eastern Beringia and with the buildup of continental ice sheets to the east. These intervals include 120,000–110,000, 75,000–60,000, and 25,000–18,000 yr B.P.

It is also interesting to note how changes in sea surface temperatures in the Beringian region lag behind the changes in insolation by several thousand years. This record is inferred from changes in marine microfossils from sediments in the northern Pacific. Intuitively, it makes sense that ocean warming and cooling should lag behind changes in air temperature because it takes far more energy to change the temperature of water than that of air, and there are untold billions of gallons of water in the Pacific Ocean. Figure II.3 also shows changes in global sea level through time. For the last 100,000+ years, sea level has apparently been below the modern level, reaching its lowest point during the late Wisconsin Glaciation (about 20,000 yr B.P.). The sea level curve is a measure of the amount of the world's water frozen in ice, so each dip in the sea level curve represents a bulge in the global ice curve.

One other feature of the Alaskan fossil record needs to be highlighted. The preservation of Pleistocene fossils in Alaska, as in other very cold regions, is made even more interesting because of permafrost. The soils in much of Alaska are permanently frozen just a few centimeters below the surface (Fig. II.4). Alaskan regions that are south of the permafrost zone today were mostly in permafrost during the Pleistocene. Dramatic evidence of permanently frozen ground is dis-

Figure II.6. Drawing of the baby mammoth, *Dima,* from the Siberian Pleistocene. (After Guthrie, *Frozen Fauna of the Mammoth Steppe,* ©1990 University of Chicago Press, reprinted by permission.)

played in such features as ice wedges (Fig. II.5). Permanently frozen sediments act as an enormous deep freeze for fossils; thus fossil specimens from frozen sediments are incredibly well preserved. Perhaps the most remarkable examples of Pleistocene fossils from permanently frozen sediments are the freeze-dried carcasses of mammals. Many of the extinct mammal species that once roamed Beringia have been found virtually intact in Alaska and Siberia. One of these was the carcass of a juvenile woolly mammoth, named *Dima* by the scientists from St. Petersburg who brought it back from Siberia (Fig. II.6). *Dima* apparently became trapped in a small, silt-choked pool on an autumn afternoon about 33,000 years ago. Sometime after the animal died, the pool froze, encasing the mammoth in frozen sediments.

This remarkable preservation of an intact animal was not an isolated occurrence, however. So many mammoths were frozen into sediments on Siberian riverbanks that tons of ivory from their tusks were hacked off the frozen carcasses or skeletons and shipped out by nineteenth-century Russian traders to be made into piano keys, buttons, and billiard balls. Victorian gentlemen were probably unaware that they were playing billiards with pieces of 30,000-year-old Siberian mammoth.

So this region has a unique fossil record preserved in a natural deep freeze. Let's move ahead and explore the paleoecology of Beringia, especially the regions that are now National Parks.

5

DENALI NATIONAL PARK

Where the Ice Age Never Ends

The Alaska Range is the backbone of the state, dividing the warmer, moister South from the cold, dry interior. Denali National Park is the heart of the Alaska Range, encompassing Mount McKinley, the highest peak of the range, as well as many other very tall mountains (Fig. 5.1). Mount McKinley (called *Denali* by the local Athapaskan Indians) is 6194 m high (20,320 ft). Nearby Mount Foraker, at 5303 m (17,400 ft) elevation, is the second highest peak in the United States. Several other mountains in the park range from 3500 to 4000 m (about 11,000–13,000 ft) in elevation.

Modern Setting

The mountains of the Alaska Range cover roughly the southern half of the park (Fig. 5.2). The regional climate, both past and present, is greatly affected by the mountains. It is said that Mount McKinley creates its own climate. A massive object jutting 6 km into the atmosphere, McKinley catches a great deal of the moisture moving north from the Pacific. Thus, the south side of McKinley and adjacent mountains receive enough moisture to generate a great deal of precip-

Figure 5.1. Mount McKinley, as seen from Wonder Lake. (Photograph by Dr. Christopher Waythomas, reprinted by permission of Dr. Waythomas.)

itation. Because the air at these high elevations is always cold, nearly all of that moisture falls as snow, which mantles the alpine landscapes in the park. Consequently, the glaciers formed on the southern flanks of the mountains are many times larger than those on the north side. This also held true in the Pleistocene.

Denali Park is in the subarctic zone. Regions below about 820 m (2700 ft) are swathed in boreal forest. The Russian term for this type of forest is *taiga,* meaning "land of little sticks." This is an apt description, as the black and white spruce trees that grow this far north are rather puny, compared with the tall trees in more southerly climes. Other trees in the boreal forest include quaking aspen, balsam poplar, and paper birch. These deciduous trees grow along streams and in places where the soil has been disturbed by fire, avalanche, or rock slide. Meadows and other open ground in the forest are generally covered with herbs, mosses, lichens, and birch and willow shrubs. Ponds in the forest zone are mostly choked with sphagnum mosses, sedges, and other aquatic and semiaquatic vegetation that is the favorite food of moose. These swampy water holes are called muskeg bogs. The filling in of a pond with vegetation and sediment may take more than a thousand years. The layers of organic detritus preserved in old, infilled bogs are a valuable source of Quaternary fossils. Above 820 m elevation, the forest gives way to tundra. Much of the park lies above **treeline.**

The tundra vegetation in Denali Park includes both moist and dry tundra plant communities, with gradations in between. Moist tundra supports the growth of herbs, such as sedges, including cottongrass. Dwarf birch and willow shrubs also grow there. Dry tundra includes well-drained soils supporting dryas, moss campion, and other low-profile cushion plants that hug the ground. These plants also persist in patches on fellfields, the rocky ground and thin soils found at higher elevations. They keep a low profile, avoiding the dehydrating, chilling winds that are so persistent in the mountains.

The wildlife of the park harvest foods in a variety of ways and in different habitats. Moose prefer the vegetation in and around ponds and muskeg bogs. They

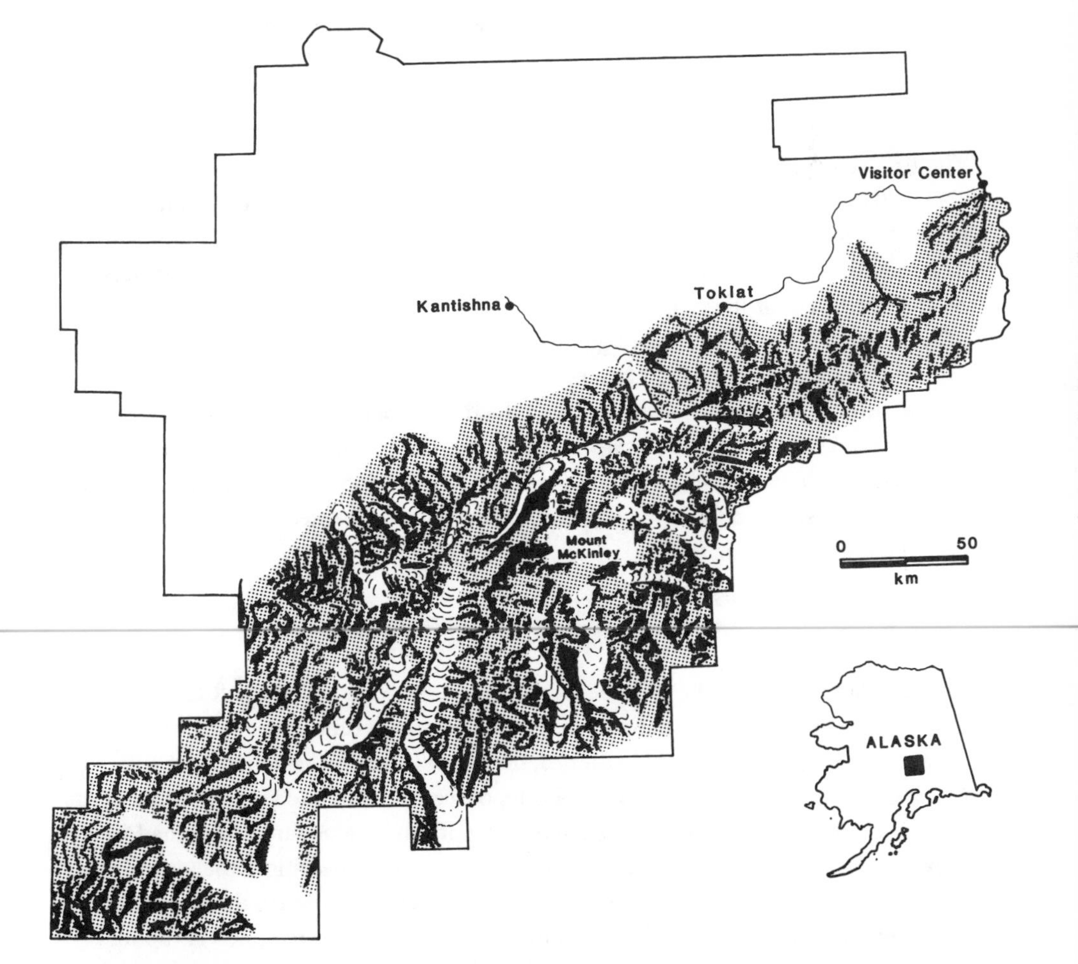

Figure 5.2. Map of Denali National Park and Preserve, showing the Alaska Range and major glaciers.

generally stay below treeline. Caribou are frequently found on the tundra. Dall sheep are nearly always seen above treeline and often frequent mountain crags and ledges, where wolves and grizzly bears cannot follow.

Wolves and grizzlies pursue prey above and below treeline, although the bears are not strict carnivores. One of their favorite prey items is arctic ground squirrel. Although the animals of Denali are considered exciting and exotic by most park visitors, they are only a small part of the truly impressive fauna that lived there only a few thousand years ago. Let's step back in time to the late Pleistocene and see what the park was like in the age of mammoths.

Glacial History: The Shaping of the Landscape

The Healy Glaciation

The first major glacial advance of the Wisconsin Glaciation took place in Alaska about 70,000 years ago. Locally, it is the called the Healy Glaciation. During that time, glaciers on the south flanks of the Alaska Range built into great ice fields and came together to form an ice sheet that covered much of southern Alaska. Glaciers on the north side of the range expanded but flowed down valleys only a few tens of kilometers; they did not come together into a regional ice sheet. One of the more important northward flowing glaciers during that time was an arm of the Nenana Glacier, which gouged the valley through which the Nenana River now flows.

The **terminal moraine** of the Healy Glaciation, situated near the town of Healy, dammed the Nenana River, forming prehistoric Lake Moody. The lake gradually filled with sediment, and the river found a new course: the steep-walled Nenana River Gorge just outside the park entrance.

The Riley Creek Glaciation

The last glacial event of the Pleistocene in central Alaska is called the Riley Creek Glaciation. This regional glaciation took place at the same time as the last major ice advance in other parts of North America, the late Wisconsin Glaciation. The Riley Creek Glaciation encompasses four episodes of ice advances and retreats between 25,000 and 9500 years ago. The glaciers did not advance as far as they had in previous glaciations. For instance, during the Riley Creek Glaciation, the terminal moraine of the Nenana Glacier extended only to the creek for which the glaciation is named, which is near Denali Park Depot. The glacier has now retreated far up the valley (Fig. 5.3). The glacial advances and retreats played a major role in shaping the landscapes of Denali.

Figure 5.3. The Nenana Glacier in 1979. (Photograph by Dr. Christopher Waythomas, reprinted by permission of Dr. Waythomas.)

Glacial Landscapes of Denali

If you look around Denali Park today, you will see a landscape that owes much of its form to the actions of glaciers. The Muldrow Glacier (Fig. 5.4), the largest of the remnants of the Pleistocene glaciers, is easily visible from the Eielson Visitor Center. However, it looks more like a low, dark mountain than a glacier because the ice is mantled by a thick load of dark debris. The U-shaped valley that lies upstream from the bridge at Savage River was carved by glacial ice, which scoured the valley sides as it crept downslope in the Pleistocene. The steep-walled valleys you see perched on the upper mountain slopes are called *cirques.* These were also sculpted by glacial ice. Unlike most of their counterparts in the Rocky Mountains, many of the cirques in the Alaska Range are still filled with ice and perennial snow.

Even the appearance of the tall peaks is due in large part to the action of glaciers that have passed through them, cutting into slopes and trimming them into pyramid shapes. The sharp angles and sheer rock walls were ground into shape by a combination of flowing ice and debris active over thousands of years. This is how the long, sharp mountain ridges called **aretes** were formed in the park. They were shaped when ice scraped away at the headwalls on both sides of a mountain, leaving over-steepened slopes that come to a sharp ridge at the top (Fig. 5.5). As ice

Figure 5.4. The Muldrow Glacier in 1978. (Photograph by Dr. Christopher Waythomas, reprinted by permission of Dr. Waythomas.)

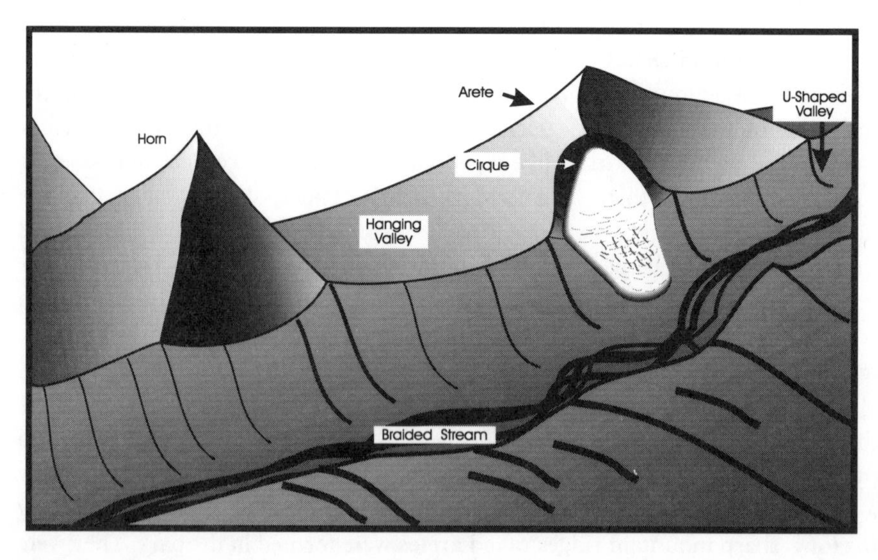

Figure 5.5. Generalized view of a glaciated mountain region in Alaska, showing topographic features carved by ice.

Figure 5.6. A lateral moraine of the Muddy Valley Glacier, Denali National Park, in 1979. (Photograph by Dr. Christopher Waythomas, reprinted by permission of Dr. Waythomas.)

flowed over different types of bedrock, it eroded the surfaces to different degrees, creating benches and stairstep valleys.

Pyramid-shaped peaks called *horns* were shaped by glaciers grinding their way past three sides of a mountain. Examples of these "Alaskan Matterhorns" include Mount Hayes and Mount Deborah in the Alaska Range east of the park.

Several types of deposits left by late Pleistocene glaciers in the Denali region are also easily visible to park visitors. First, great piles of rock, sand, and silt run out from the flanks of the Alaska Range. These are terminal and **lateral moraines** left by retreating glacial lobes (Fig. 5.6). Large boulders, called glacial erratics, were carried by glacial ice from the mountains to the foothills and then dropped when the ice melted. If you see a large boulder that looks out of place on an otherwise gently rolling tundra knoll in the park, it is probably a glacial erratic (Fig. 5.7).

The braided streams and rivers in the park are choked with glacial debris. Glacial silt, sand, and gravel accumulated in tremendous amounts as the glaciers ground their way forward; then, when the ice retreated, the debris was left behind to be moved by water and wind. The combination of ancient glacial debris and the current crop being ground up by modern glaciers mix together to fill the waters with silt, silt, and more silt. Backpackers in Denali soon learn to avoid collecting drinking water from these silty streams (or they learn to tolerate drinking gritty

Figure 5.7. A glacial erratic boulder, Denali National Park. (Photograph by Dr. Christopher Waythomas, reprinted by permission of Dr. Waythomas.)

water). During field work in 1990, we collected water from the Foraker River, let it sit for up to several days, and still couldn't get all the silt to settle to the bottom. Needless to say, silty water quickly clogs water purification filters.

The channels of those braided streams are constantly shifting, leaving old channels to dry. When this happens, the deposited silts are easily picked up by winds and carried great distances (Fig. 5.8). Much of interior Alaska is mantled with tens of meters of this silt.

The Ice Age and Steppe-tundra

What do we know of life in and around Denali during the last ice age? Both the geologic and fossil records show that it was a cold, dry, periglacial landscape, not unlike what can be seen today in the higher alpine zones of the Alaska Range. The loess that drapes the landscapes north of the Alaska Range was periodically colonized by plants during the Healy and Riley Creek glaciations. Depressions on the surface filled with meltwater during summer. Some depressions became ponds and lakes that accumulated organic-rich sediments. The fossils from these sediments provide the main record of regional life during the ice age.

Evidence from Organic Deposits

The mixture of partially decomposed organic detritus and silt became frozen into successive layers through many hundreds or thousands of years. This curious mixture is called "muck" by Alaskan miners, anxious to melt it out and strip it away to get at the placer gold deposits underneath. In the process of blasting away with water hoses at hillsides of frozen muck, miners have uncovered numerous Pleistocene mammal carcasses, such as the extinct large-horned bison, *Blue Babe* (Fig. 5.9), which lived about 200 km (120 miles) north of the Denali region about 35,000 years ago.

Layers of muck have also been melted out in natural exposures, such as riverbanks. I have studied organic deposits from this type of exposure along the Foraker River in the park (Fig. 5.10). The organic detritus in the frozen sediments does not fully decompose until it thaws, so it is in a strange state of suspended animation. When the layers are exposed, after perhaps 50,000 years in the frozen ground, their fossils start decomposing again. A sickly sweet stench fills the air at these exposures. The fossils, smelly or not, have an intriguing story to tell about ice-age environments and biotic communities.

Figure 5.11 summarizes the events of the last glacial cycle in the Alaskan interior. The Healy Glaciation saw the expansion of ice, especially to the south of the Alaska Range. On the north side, cold, dry conditions prevailed, supporting the

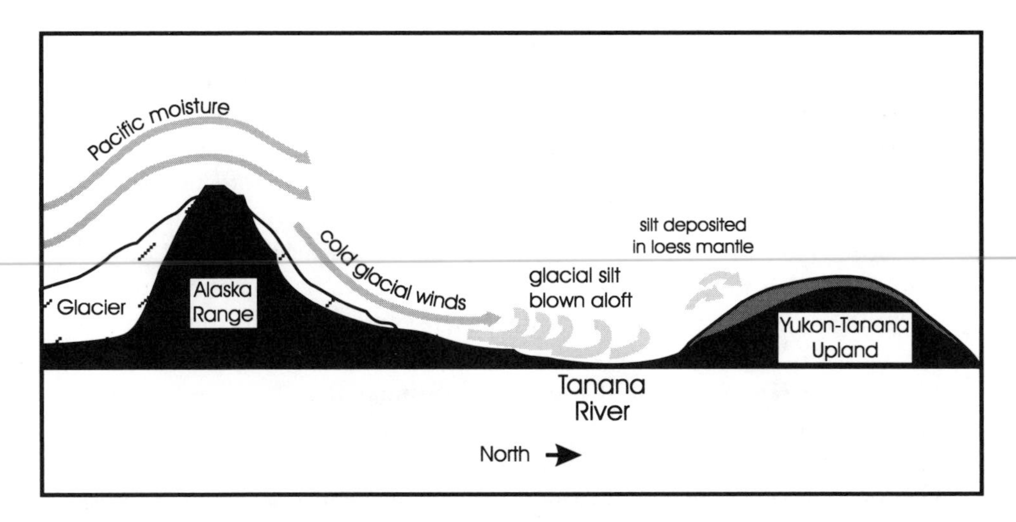

Figure 5.8. Origin of the loess mantle that covers much of the Alaskan interior north of the Alaska Range. Water carried glacial silt to the major river deltas, including the Tanana. Clouds of silty dust were blown aloft and then deposited on regional uplands. (Modified from Guthrie, 1990.)

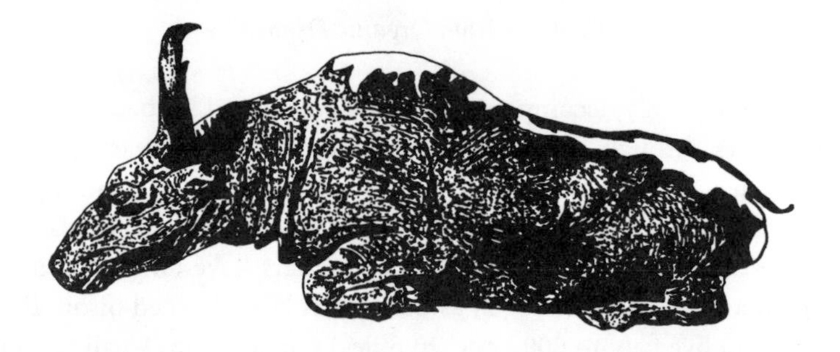

Figure 5.9. Pleistocene large-horned bison mummy, found in frozen "muck" north of Fairbanks by a placer gold miner. The bison, named "Blue Babe" because of a blue mineral deposit that encrusted its hide, is on display at the University of Alaska Museum in Fairbanks.

development of steppe-tundra. This ecosystem, unique to the Pleistocene, stretched from western Europe across Asia and the Bering Land Bridge to Alaska and the Yukon Territory (Fig. 5.12). It was more pervasive during glacial intervals because the cold, dry conditions allowed grasses and other drought-resistant plants to outcompete other types of vegetation.

The Nature of Pleistocene Steppe-Tundra

Although it is hard to imagine, the steppe-tundra was a highly productive ecosystem. It has been admirably described by paleontologist Dale Guthrie in his book *Frozen Fauna of the Mammoth Steppe.* Unlike the modern-day Arctic, where soils are often very wet, and the active layer over permafrost is shallow, the soils of the Pleistocene steppe-tundra were drier and thawed more deeply in summer. These soils yielded their nutrients more readily to plants. The steppe grasses of central Asia were preadapted to these conditions, and during glacial intervals of the Pleistocene they spread across Eurasia and Beringia, combining with tundra vegetation (plants adapted to moister conditions in cold climates). The combination of steppe and tundra plant species formed a rich mosaic of vegetation, supporting a splendid mammalian fauna (Fig. 5.13). The only modern ecosystem that supports a similar variety of grassland animals is the African savannah. Interestingly, the Pleistocene steppe-tundra had several mammalian counterparts to the modern African fauna, including woolly mammoth (elephant), woolly rhino (rhinoceros), and saiga antelope (antelope). Instead of zebras, the steppe-tundra had Pleistocene horses. Instead of water buffalo and wildebeest, the steppe-tundra had large-horned bison and two species of muskox. The dominant grazers on this cold grassland were the woolly mammoths, Pleistocene horses, and large-horned bison.

Figure 5.10. Organic deposits from the last glaciation, exposed in ground ice formation along the Foraker River, Denali National Park.

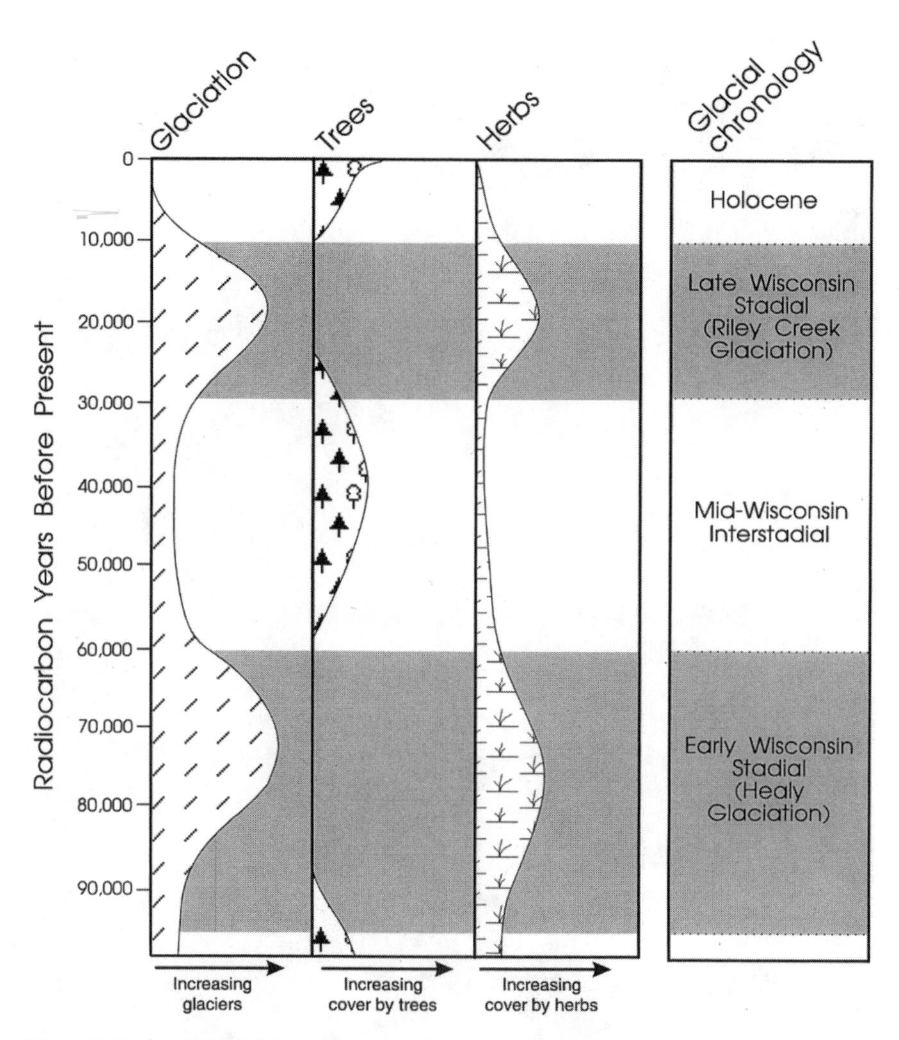

Figure 5.11. Summary of glacial events and changes in regional biotic communities in the Alaskan interior during the Wisconsin glaciation. Glaciers in the Alaska Range grew in two major cold episodes, or stadials, during which steppe-tundra vegetation was dominant on unglaciated landscapes. A mid-Wisconsin warming, or interstadial, saw an expansion of spruce forests. (Modified from Guthrie, 1990.)

These herbivores, along with the grazers that still live in Alaska, were preyed on by a variety of predators. The Pleistocene lion looked very much like modern lions but was larger (it had to deal with some very large prey animals). The saber-toothed cat was also a powerful predator, armed with daggerlike canine teeth for slicing into the neck of prey animals. However, as far as I am concerned, the most impressive Pleistocene predator was the giant short-faced bear. This animal made grizzly bears look puny (Fig. 5.14). It stood about 2 m tall at the shoulder (6 ft) and was 3 m from nose to stubby tail (10 ft). Its long legs were adapted for running down prey. We know that grizzlies, even with their relatively short legs, can put on bursts of speed of up to 65 km (40 miles) per hour. Giant short-faced bears undoubtedly ran faster than that, and they probably were able to sustain their speed over greater distances. We have no evidence that these bears sought humans as prey, but some anthropologists have speculated, only half jokingly, that the New World was not a safe place for people to live until the short-faced bear died out, about 10,000 years ago.

Research in other regions of Alaska has shown that steppe-tundra did not cover all of Eastern Beringia. Rather, there was a mosaic of ecosystems ranging from fellfield tundra on the coldest, driest slopes to bogs and fens in wet lowlands (Fig. 5.15). In southwestern Alaska, my research team from the University of Colorado has discovered that mesic tundra dominated most landscapes even during the coldest, driest intervals of the last glaciation. Mesic tundra is composed of sedges, moisture-loving grasses and other herbs, cushion plants, and shrub birch and willow. This is the dominant type of tundra vegetation found in arctic Alaska today. Undoubtedly, other regions in Alaska also supported various types of tundra vegetation during the last glacial cycle. We get only glimpses of past ecosystems through their fossil records, so our view of ice-age Beringia remains rather limited. However, the diversity of mammals found in the Alaskan fossil record is in itself a strong argument that there were many types of habitats on the prehistoric landscape.

Between the Glaciations: Spruces Rebound

The Healy Glaciation ended about 60,000 years ago. It was followed by an **interstadial,** a long period of milder climate, named the Boutellier Interstadial by

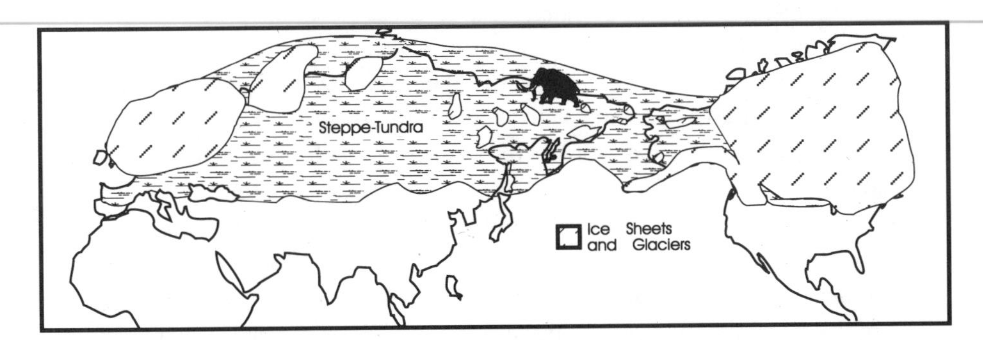

Figure 5.12. Map of the Northern Hemisphere during the Wisconsin Glaciation, showing extent of ice sheets and the steppe-tundra ecosystem. (Modified from Guthrie, 1990.)

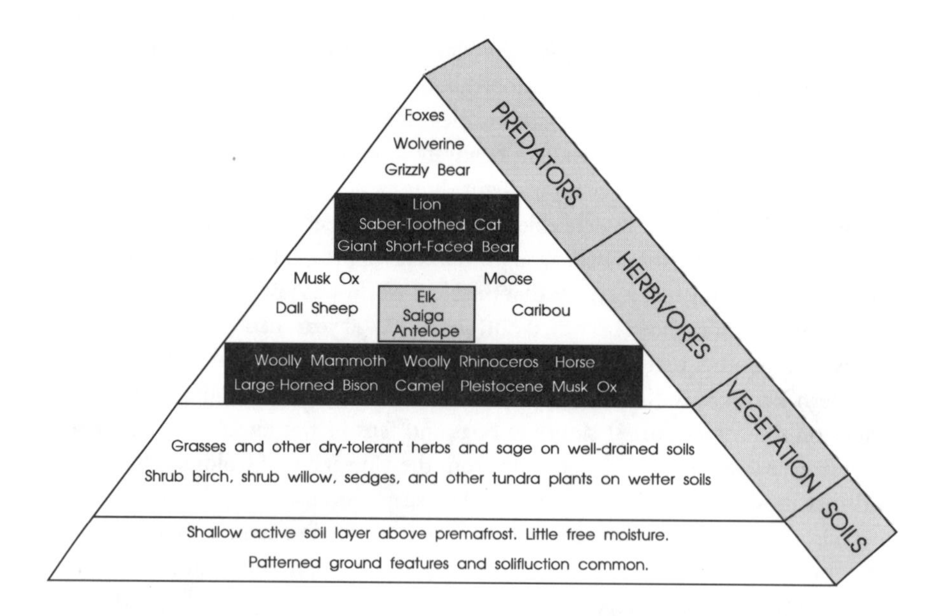

Figure 5.13. Summary diagram of the steppe-tundra ecosystem, showing the principal predators and herbivores, dominant vegetation types, and soil conditions. Species in the black boxes are extinct. Species in the gray box are no longer native to Alaska.

Quaternary geologist David Hopkins (based on interstadial deposits described from Boutellier Pass, on the Alaskan Highway in the western Yukon Territory).

The Boutellier Interstadial

During the Boutellier Interstadial, climate warmed sufficiently to cause glaciers to retreat but did not achieve full interglacial levels of warming (i.e., it was not as warm as it is today). We know this is the case in Denali National Park because our recent research project there, along with other regional studies, found that the vegetation and insects from interstadial deposits represent a mixture of boreal and tundra species. Spruces, alders, and tree birches reestablished themselves in the Alaskan interior during this interstadial, but they did not form a dense forest. Rather, they existed in an open **parkland,** sharing the landscape with patches of tundra vegetation.

Similar mixtures of forest and tundra communities exist today in regions at the northern edge of the boreal forest in Canada, although they do not match interstadial communities exactly because the environmental conditions under which the interstadial communities formed have no exact modern counterpart. Curiously, even though the Boutellier Interstadial lasted 30,000 years, spruces were unable to become established in some other regions of Beringia, notably in southwestern Alaska. That region remained covered with shrub tundra (dominated by birch and willow shrubs) throughout the Boutellier. It was apparently at the far end of the migration route of conifers into Alaska from the south and east, along the Yukon River drainage, and the trees simply did not make it into this region before the next glacial period began.

The steppe-tundra ecosystem did not disappear entirely during the Boutellier. It persisted in many regions and was able to spread southward to the southern boundary of the Cordilleran (western) and Laurentide (eastern) ice sheets. This was made possible by the development of an ice-free corridor between the two big ice sheets (Fig. 5.16). That ice-free corridor likely played an important role in the dispersal of human populations from Beringia to regions south of the continental ice sheets when it opened up again, near the end of the Wisconsin Glaciation.

Extinctions at the End of the Pleistocene

The Riley Creek glacial advance was the "last hurrah" for the Pleistocene glaciers of the Alaska Range. It was brought on by climatic cooling that was well developed by 25,000 years ago. Once again, forests retreated, and steppe-tundra expanded to cover much of unglaciated Beringia. This cold episode also turned out to be the swan song for a substantial number of large mammals. All the proboscidians (mammoths and mastodons), the horses, camels, and giant sloths became extinct

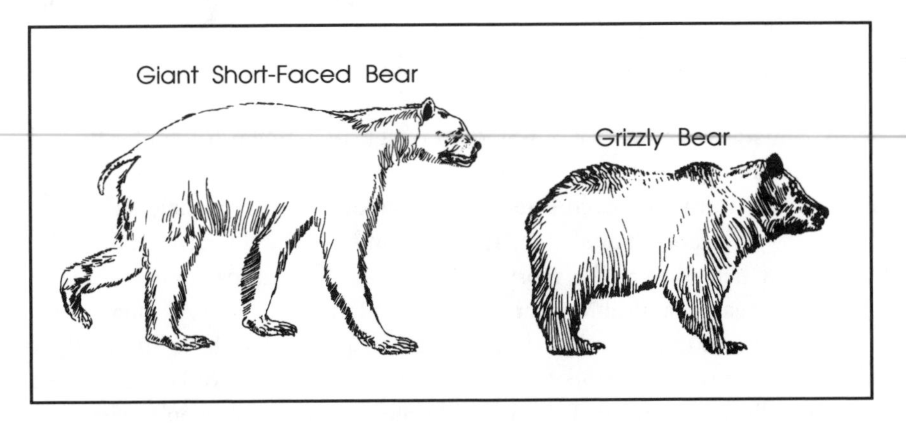

Figure 5.14. Size comparison between the giant short-faced bear and modern grizzly bear. (After Ruth Ann Border, *Mammoth Graveyard: A Treasure Trove of Clues to the Past,* 1988, Hot Springs, SD, The Mammoth Site.)

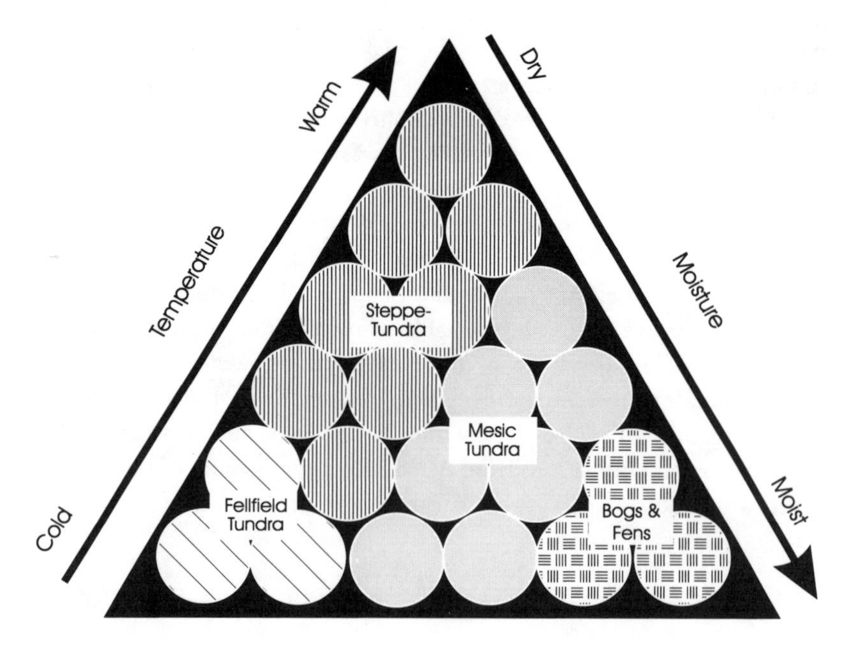

Figure 5.15. Diagram showing mosaic of vegetation types associated with various temperature and moisture conditions in Alaska during the late Pleistocene.

in North America at the end of the last glaciation. Why did this happen? The obvious answer might seem to be that these cold-adapted animals could not tolerate the warm climates of the Holocene. This might be convincing if it weren't for the fact that the same cold-adapted species and their ancestors had withstood the warm climates of about a dozen previous interglacial periods, at least one of which was probably substantially warmer than anything yet experienced in the Holocene.

No, climatic warming per se was not enough to extinguish the **megafaunal mammals** (animals with adult live weight greater than 40 kg are termed megafauna). There must have been some unique biological or environmental factors influencing megafaunal populations at the beginning of the Holocene. Some argue that human beings were the most important agent in dispatching the North American megafauna. Paul Martin, geologist at the University of Arizona, coined the phrase "Pleistocene overkill" to describe how people may have hunted these animals to extinction. The theory suggests that the megafauna of North America was especially vulnerable to Paleoindian hunters because the people were newcomers on this continent at the end of the Pleistocene, and the animals, unaccustomed to human hunters, had little natural fear of them and were thus too easily hunted. The theory holds that this new hunting pressure, combined with rapid

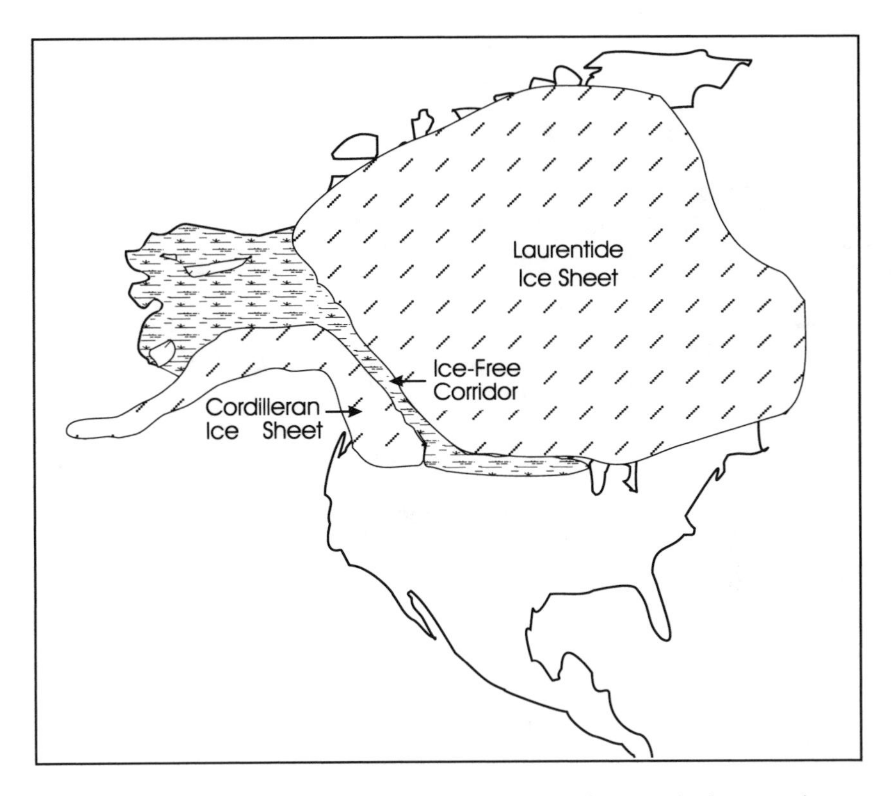

Figure 5.16. Sketch map showing establishment of an ice-free corridor between the Laurentide and Cordilleran ice sheets during the Mid-Wisconsin Interstadial (ice positions approximate only). This allowed the southward spread of steppe-tundra species to regions south of the ice sheets.

climate change, wiped out most of the megafauna on this continent. We will probably never be able to confirm the overkill theory, since the fossil evidence is so spotty. Whatever the cause of the demise of North American megafauna, we are left with a rather sparse collection of large animals that have made it through the Holocene. Nevertheless, they are true survivors. If they were human, they would probably sport hats and T-shirts with the motto, "We survived the Pleistocene–Holocene transition!" (Fig. 5.17).

Development of the Modern Ecosystems of Denali

Because of changes in Earth's orbit around the sun (the Milankovitch cycles), climates began to warm significantly in the Northern Hemisphere, starting about

Figure 5.17. Some of the megafaunal mammals that survived the Pleistocene–Holocene transition in North America.

13,000 years ago. The first indication of this warming in Alaska was when steppe-tundra vegetation and insects adapted to cold, dry conditions were replaced by birch shrub tundra and insects adapted to warmer climates. When the birch shrubs had gotten thoroughly established, they began to alter the nutrient levels and chemistry of Alaskan soils, allowing alders to become dominant. By about 9000 years ago, spruce invaded east-central Alaska. Both black and white spruces spread rapidly across interior Alaska, north of the Alaska Range. In the Denali region, the modern ecosystems were pretty much established by about 6000 years ago.

Suggested Reading

Bartlein, P. J., Anderson, P. M., Edwards, M. E., and McDowell, P. F. 1991. A framework for interpreting paleoclimatic variations in Eastern Beringia. *Quaternary International* 10–12: 73–83.

Collier, M. 1989. *The Geology of Denali National Park.* Anchorage: Alaska Natural History Association. 48 pp.

Guthrie, R. D. 1990. *Frozen Fauna of the Mammoth Steppe. The Story of Blue Babe.* Chicago: University of Chicago Press. 323 pp.

Hopkins, D. M., Matthews, J. V., Jr., Schweger, C. E., and Young, S. B. 1982. *Paleoecology of Beringia.* New York: Academic Press. 489 pp.

Johnson, P. R., and Hartman, C. W. 1969. *Environmental Atlas of Alaska.* College, Alaska: University of Alaska Press, 111 pp.

6

BERING LAND BRIDGE NATIONAL PRESERVE

Gateway to a Vanished Ecosystem

Recently, the National Park Service established the Bering Land Bridge National Preserve on the Alaskan side of the Bering Strait. This large region covers much of the northern part of the Seward Peninsula (Fig. 6.1). It safeguards a part of a modern coastal tundra ecosystem from further development, and it commemorates events that took place there thousands of years ago. It is not unlikely that the ancestors of the native Americans took their first steps on the North American continent in or near the preserve.

The Bering Land Bridge is probably the single most important piece in the puzzle of how human beings first arrived on the North American continent. We have already seen how the land bridge acted as a conduit for steppe-tundra plants and animals in the Pleistocene. Now we will take up the threads of another story: the peopling of the New World. On the basis of archaeological evidence, we can be reasonably certain that the people who first made the trip into North America were hunters following game animals (Fig. 6.2). Those hunters, called "Paleoindians" (ancient Indians) by anthropologists, were intimately involved in the natural world around them. They were living in a harsh environment, relying on their knowledge of weather, terrain, and animal behavior as they staked their survival on hunting the fauna of the steppe-tundra. Let's look at the natural world they lived

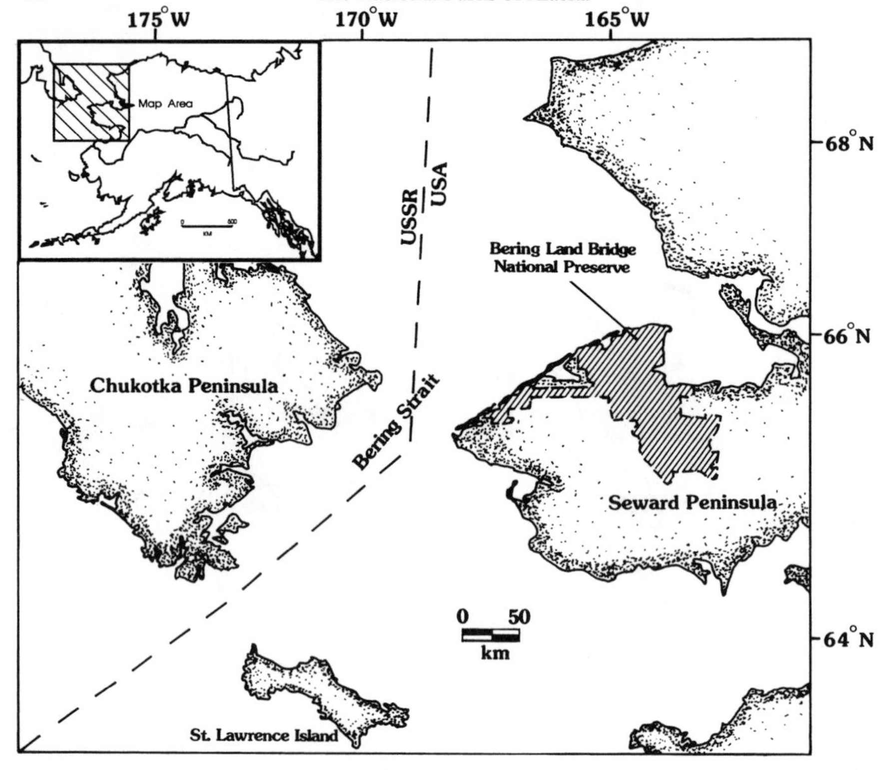

Figure 6.1. Map of western Alaska and eastern Siberia, showing location of Bering Land Bridge Preserve.

in, as they made their sojourn into a new land, probably some time before 12,000 years ago.

Formation of the Bering Land Bridge

The continental shelves between Alaska and Siberia were dry land during the Wisconsin glacial and interstadial intervals, connecting Alaska and Siberia across a broad front. We know this from several lines of research. First, there is evidence from around the world, including Alaska and Siberia, that sea level was lower during Pleistocene glaciations. Part of the water that would normally have contributed to filling the world's seas was caught up in continental ice sheets and glaciers. The continental shelves between Alaska and Siberia are unusually shallow, so it follows that they were above sea level whenever sea level was substantially lower.

We also have evidence from the deposits that accumulated on the land bridge when it was dry land. Geologists at the U.S. Geological Survey have taken sediment cores from the continental shelves off the western coast of Alaska and discovered layers of peat and other sediments that accumulated on the land bridge. These layers lie beneath marine sands that swept over the land bridge when sea level rose and saltwater flooded the continental shelves. In some regions, peaty land-bridge deposits are still visible beneath the shallow waters of the Bering and Chukchi Seas. I have this on the authority of an Inuit student in the audience of one of my lectures, who informed me that she and her family had seen the peat deposits I was talking about as they boated across the Chukchi Sea to visit some friends on the Siberian coast.

Finally, geologists have used sonar to map the surface features of the now-submerged land bridge. The sonar mapping is based on sound waves, sent in pulses from a ship, which reflect off the surface of objects in the water (Fig. 6.3). They have found traces of ancient stream beds, deltas, and even glacial moraines that formed on dry land many thousands of years ago.

Land Bridge Environments

At the northern edge of the Bering Land Bridge, the continental shelf of the Arctic Ocean was also above sea level. The Arctic Ocean was cut off from warm Pacific water when the land bridge came into being. The Arctic Ocean froze over and stayed frozen at the surface until the land bridge was inundated at the end of the last glaciation. The northern coast of the land bridge was probably extremely cold and offered only marginal habitats for terrestrial life. The vegetation in this region was likely similar to that found in modern **polar deserts** of the high arctic. Because

Figure 6.2. Artist's impression of Pleistocene hunters following woolly mammoths across the Bering Land Bridge under the glow of the Aurora Borealis. (Modified from the logo of the Center for the Study of the First Americans, Oregon State University, Corvallis, Oregon, reprinted by permission.)

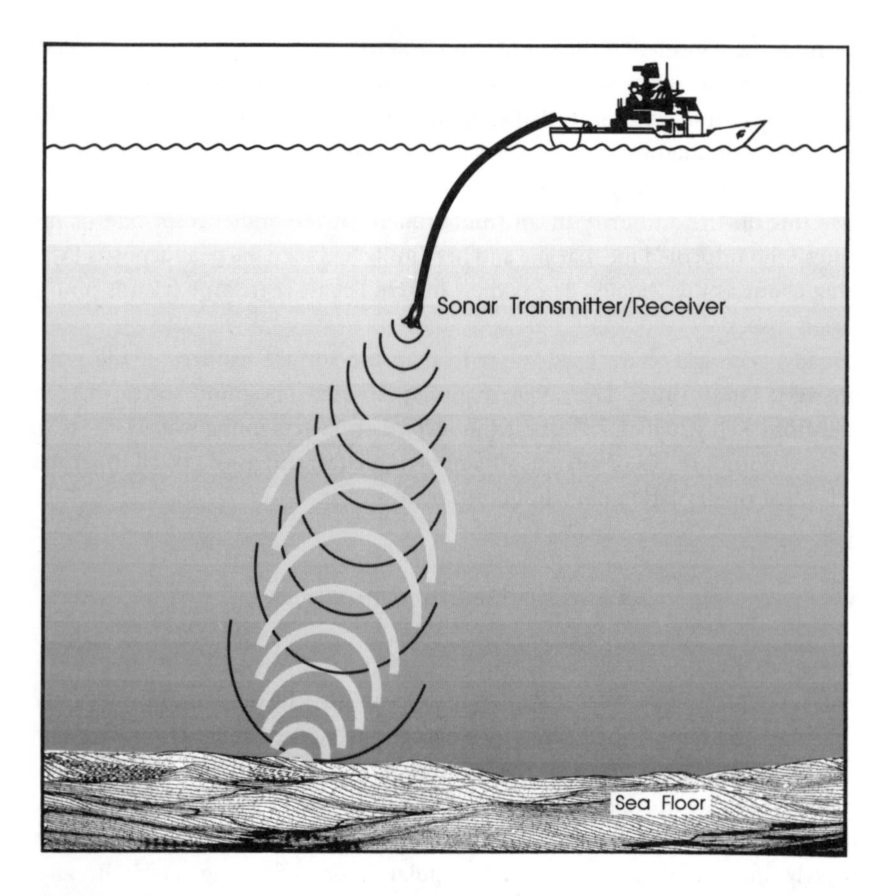

Figure 6.3. Mapping of sea floor terrain by side-scan sonar. Sound waves are emitted from a transmitter towed behind a boat. The waves bounce off the surface of the sea floor and then are picked up by the receiver. The pattern and intensity of sound waves received are translated into digitized data. When these data are compiled for a whole region, a topographic map of the sea floor can be constructed.

the northern coast of the land bridge was ice-bound, it probably supported few marine mammals for Paleoindians to hunt.

The cutting off of Pacific waters and moist air masses by vast expanses of continental shelf at the southern end of the land bridge produced the cold, dry **continental climate** that fostered the steppe-tundra biome across Beringia. We do not have a clear picture of what the vegetation was like on much of the land bridge itself, however, because so little work has been done on sampling and studying fossil pollen and macrofossils from land bridge sediments. The soils of the land bridge developed from marine sediments that were exposed to the air when sea level dropped. Wind-blown loess from adjacent Siberia and Alaska may have contributed to the buildup of terrestrial soils. It is extremely difficult to say what these soils were like, because there is nothing to compare them with in the modern world. About the only modern parallel might be the Dutch polders, the land reclaimed from the sea in the Netherlands. However, the Dutch have put considerable effort into building up those soils in order to make them suitable for farming, so that comparison with the Bering Land Bridge would not be very useful.

A fossil mammoth site from the Baldwin Peninsula of Alaska (Fig. 6.4), dating to about 27,000 yr B.P., provides some information on mid-Wisconsin Interstadial environments in the central region next to the land bridge. The sediments surrounding the mammoth yielded pollen, plant macrofossils, and insects. Taken in combination, these fossils are indicative of a shrub tundra environment. There is no evidence of trees, and steppe-tundra elements are minimal. Insects, pollen, and plant macrofossils from the nearby Cape Deceit site, located just east of the boundary of the Bering Land Bridge Preserve, confirm the regional dominance of shrub tundra during mid-Wisconsin times, but during the subsequent glacial interval, they reflect a shift to cold, dry conditions and steppe-tundra.

This interpretation agrees with the late-Wisconsin-age pollen data from St. Lawrence and St. Paul Islands. Palynologist Paul Colinvaux studied fossil pollen

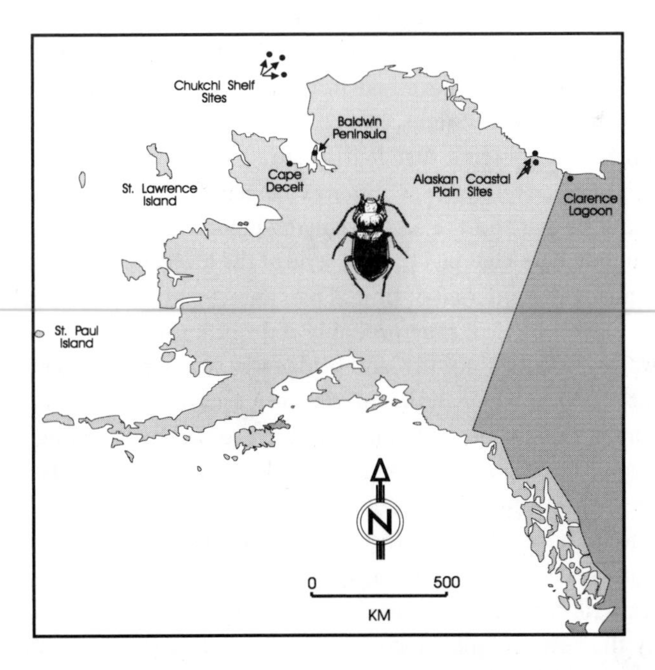

Figure 6.4. Map of Alaska, showing location of paleontological sites discussed in the text.

from lake sediments on these islands (which were highland regions on the ancient land bridge). The late Pleistocene pollen data suggested cold, dry, steppe-tundra environments until after 11,000 yr B.P. The postglacial climatic warming associated with the establishment of shrub birch around Colinvaux's study sites was apparently quite late on these islands compared with the arrival of birch in adjacent mainland regions. Apparently, major portions of the land bridge were covered with steppe-tundra during the height of the last glaciation.

On the basis of fossil evidence from adjacent Alaska and Siberia, some researchers have visualized the southern part of the land bridge as having supported a mosaic of vegetation types, including deciduous woodland, tall brush, meadows, and marsh vegetation as well as grasslands or steppe. Other than Colinvaux's work from St. Paul Island, we have no actual data from that portion of the land bridge to support this view.

As discussed in the last chapter, at the end of the last glaciation (circa 14,000 yr B.P.), the vegetation throughout Alaska began to change from steppe-tundra to mesic shrub tundra. Dwarf birch spread rapidly across the Beringian landscape. Changes in regional pollen suggest a climatic change to warmer, moister conditions in the summer and perhaps greater snowfall in winter. The birch invasion was probably tied to a dramatic regional warming at the end of the last glaciation. It was fostered by a rapid breakdown of the cold, dry climatic conditions as sea level rose and the land bridge was inundated.

This vegetation change, in conjunction with the disappearance of the land bridge under the sea, must have had immediate and profound effects on other elements of Beringian ecosystems, especially the large mammals and, presumably, human cultures. Ecosystems that had endured for tens of thousands of years vanished in a matter of decades as climate changed and sea level rose. The drowning of the land bridge and the accompanying environmental changes took place rapidly enough to have changed the life-style of the few generations of people who were there during that interval of time. Their ancestors had moved freely between Siberia and Alaska. Indeed, they probably did not know that they were moving between two continents, since the land bridge was such a broad region. This route was now cut off. Meanwhile, in the interior of Alaska and the Yukon, conifer trees began to spread rapidly along major waterways, possibly providing fuel used by Paleoindians for heating and cooking (a fuel that had been unavailable in Eastern Beringia for the previous 25,000 years).

In order to understand land-bridge environments at the time when people were venturing into Alaska, we need to know two more things: (1) How rapidly did the Beringian climate change at the end of the Pleistocene? (2) How warm did it get?

Probably the best evidence documenting the timing and intensity of that climate change comes from insect fossils. A thermophilous (warmth-loving) beetle fauna was found by my colleague John Matthews, from 10,900-year B.P. sediments at Clarence Lagoon, on the northwestern coast of the Yukon Territory (Fig. 6.4). This fauna reflects climatic conditions similar to those of today. My work on fossil beetles from the northeastern coastal plain of arctic Alaska suggests that summer temperatures were 2–3°C (4–5°F) warmer than at present by 10,400 yr B.P. I have also studied 11,000-year-old insect fossils from peat samples in cores taken from the continental shelf of the Chukchi Sea (Fig. 6.4). Those fossils show that summer temperatures were substantially warmer than conditions found on adjacent parts of arctic Alaska today! This evidence is rather startling, since it appears to contradict the Alaskan pollen evidence showing a gradual shift from steppe-tundra to shrub tundra to coniferous forest, a process that took several thousand years to complete. However, the insects were apparently able to take advantage of warming climates much more readily than their plant counterparts, because plant communities are tied to a whole spectrum of environmental conditions (especially, in this case, soil development). On the other hand, insects, especially predators, are free to go wherever the climate is suitable, so long as there are other insects to prey on.

Let me clarify here that the insects in these sites are still found in arctic and subarctic regions today. So, even though they indicate that summer temperatures were as warm as or warmer than today, that does not mean that arctic Alaska had a climate as warm as south Florida. The terms "warm" and "warmer than present," however, are *relative* terms. They mean that summer temperatures were warmer than might be expected for arctic regions at the end of an ice age. Winter conditions probably remained harsh and very cold, even if they were a few degrees warmer than they had previously been. At any rate, human beings scarcely appreciate any difference between –45°C and –55°C (–49 and –67°F). Both temperatures just feel *cold.*

The reconstruction of the timing and intensity of climatic warming based on insects also coincides with the peak of solar radiation predicted by the Milankovitch model. So, we have our answers: the climate got very warm very fast at the end of the Pleistocene. In fact, the transition happened so fast that the plant communities of Beringia had a very hard time keeping up.

Sea-level History

The other crucial factor in any discussion of the peopling of the Americas is: When was the Bering Land Bridge inundated by the sea? That inundation eliminated the land bridge as a means for people to get from the Old World to the New. It is important to realize that the Bering Land Bridge was not some narrow isthmus between Siberia and Alaska. It was a huge region, covering more than a million and a half square kilometers (579,000 square miles, about twice the size of Texas). At the height of the last glaciation, the distance from the northern margin of the land bridge to the southern margin was 1800 km (1125 miles), roughly the distance between Buffalo, New York, and Miami, Florida. So, when people crossed the land bridge, they probably had no idea that they were moving from one continent to another. They were simply traversing a huge lowland area, and they kept moving east until they unwittingly entered North America.

Sea-level reconstructions based on radiocarbon dates on ocean sediments from the continental shelves suggested that the land bridge was inundated by about 14,500 yr B.P. It turns out that those ^{14}C dates were probably several thousand years too old, because the carbon in the test samples was contaminated with coal. The ancient carbon in coal is devoid of ^{14}C, so it dilutes the concentration of radiocarbon in a sample, making it appear older than it really is. Our research group at the University of Colorado has now published radiocarbon ages on peat macrofossils from the Chukchi Shelf. Our data suggest that the inundation was not completed until after 11,000 yr B.P. and that substantial portions of the land bridge remained above water until that time. This means that Paleoindian hunters could have walked across the land bridge until then. Furthermore, the inundation of the land bridge coincided closely with the warmer-than-modern climatic episode shown by the insect fossils rather than predating it by several thousand years, as the previous sea-level reconstruction had suggested. The land bridge itself probably acted to strengthen the hold of cold, dry climates on Eastern Beringia during the late Pleistocene. It blocked the oceanic circulation between the Pacific and Arctic Ocean waters, and warm, moist air from the Pacific was separated by more than 1000 km (625 miles) from the center of the land bridge and adjacent continental regions.

As the land bridge was inundated, the reestablishment of circulation between the Pacific and Arctic Oceans through the Bering Strait probably played a major role in climatic warming in Eastern Beringia around 11,000 yr B.P. Our evidence from the land bridge suggests that this warming was both rapid and intense.

Beringia and the Peopling of the Americas

One pitfall in our thinking about people crossing the land bridge is the notion that it was impossible for people to get across the Bering Strait region after the land bridge was inundated. In fact, people are still walking across the Bering Strait today—in winter. Ice frequently covers the entire surface of the Bering Sea between the Seward Peninsula in Alaska and the Chukotka Peninsula in Siberia. Inuit villagers from the two regions visit each other by simply walking (or snowmobiling) across the frozen sea.

Another option for prehistoric peoples would have been to traverse the Bering Strait by boat. We have abundant evidence that many prehistoric cultures built seaworthy vessels and traveled great distances in them. That is how the Polynesian people made their way to the Pacific islands from New Zealand. Archaeologist James Dixon of the Denver Museum of Natural History has reviewed the evidence for the peopling of the New World, including recent discoveries of early habitation in southern South America, and he has proposed a maritime scenario that would allow rapid movement of peoples from the Kamchatka Peninsula of Siberia along the Pacific coast as far south as Chile. By boat, this migration may have taken only a few decades or, at most, a few centuries (Fig. 6.5).

This migration theory takes into account the discoveries of archaeologist Tom Dillehay, who recently made a startling and controversial breakthrough in the field of Paleoindian archaeology with his research at a site called Monte Verde, near the Andes Mountains in Chile. Dillehay found the waterlogged remains of a Paleo-indian camp that had been buried by peat layers and preserved nearly intact. The ancient land surface, preserved under younger layers of sediment, remained essentially undisturbed and showed dramatic evidence of human habitation, including such features as post holes and hearths. The site also contained a wealth of artifacts, which include far more than the typical stone tools. Because of the unusual preservation in a waterlogged environment, the Monte Verde site preserved wooden tools, leather goods, seeds, and other food remains, which have been radiocarbon dated from 13,000–12,500 yr B.P. These artifacts predate the earliest reliable Paleoindian sites in Alaska as well as the Clovis culture sites throughout North America (the Clovis culture is the earliest recognized Paleoindian culture in the Americas, named after a type of **projectile point** discovered near Clovis, New Mexico). How did these people get to Chile seemingly before most of the rest of the New World was colonized? Dixon's answer is that they came by boat.

He postulates that, about 14,000 yr B.P., Asiatic peoples living on the Pacific Rim (which then extended only as far north as the southern margin of the Bering Land Bridge) traveled by boat, eastward along the land bridge to the Pacific coasts of North and then South America, reaching southern Chile before 13,000 yr B.P. These people had a maritime culture, specializing in harvesting fish and marine mammals in coastal environments. Any camp sites they might have made along the west coast of North America would now be under water, as sea level rose at the end of the Pleistocene, covering the continental shelves that were exposed during late Wisconsin time. Groups of seafarers may have traveled inland in North America, founding the Clovis culture of the American Southwest, beginning about 11,800 yr B.P.

The most widely held alternative theory among archaeologists is that the New World was first entered on foot over the Bering Land Bridge. However, archaeologists were slow to discover sites relating to these first inhabitants of the New World.

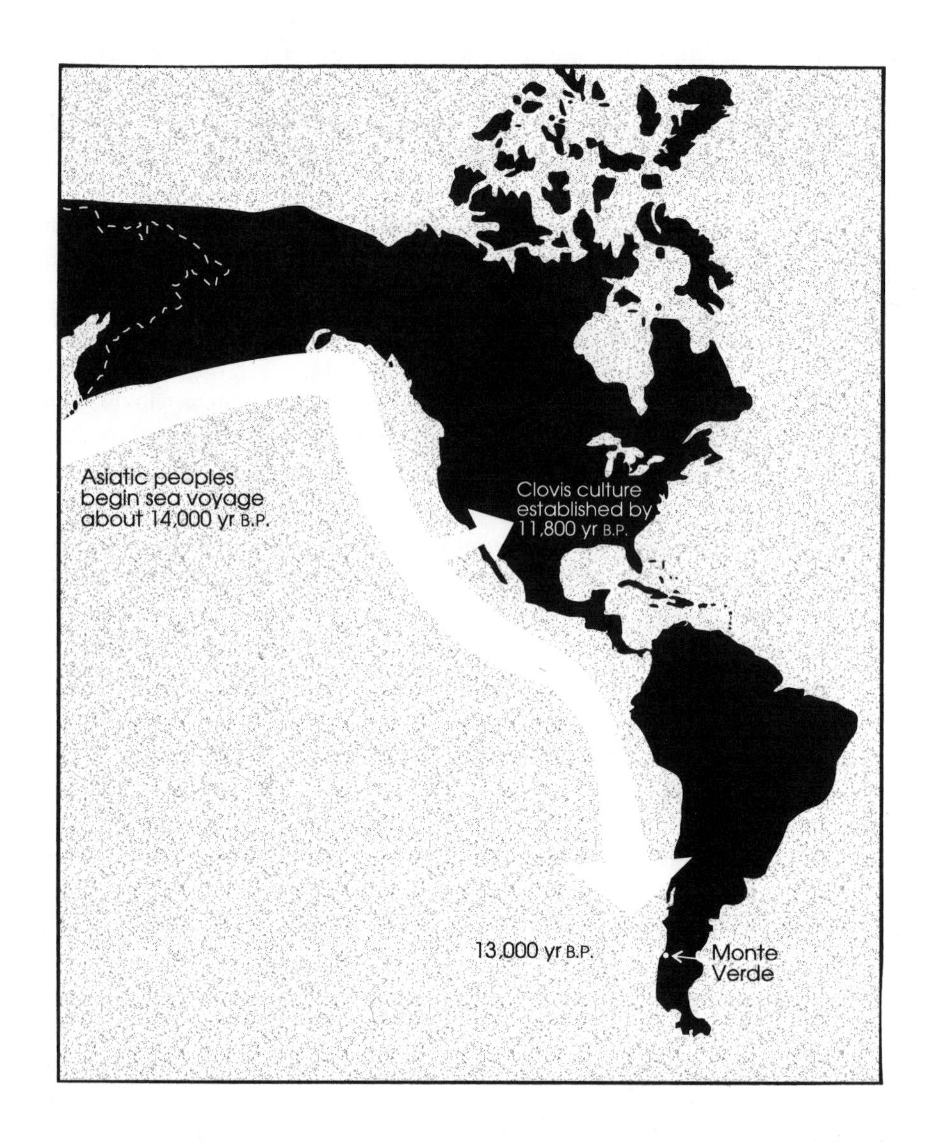

Figure 6.5. Map depicting theory of maritime migration of people from Asia into the New World, beginning about 14,000 years B.P. (After E. James Dixon, *Quest for the Origins of the First Americans,* © 1993, University of New Mexico Press, reprinted by permission.)

Work in the 1960s and 1970s finally revealed Paleoindian artifacts in Alaska and the Yukon, but the artifacts, stone tools, were not what was expected. Instead of matching the earliest stone tools found elsewhere in North America, the projectile points of the Clovis culture, these tools resembled those found in late Pleistocene sites from northeastern Asia.

Recently, archaeologists Michael Kunz and Richard Reanier discovered an early occupation site on a mesa on the north slope of Alaska (Fig. 6.6). The site contained projectile points and charcoal, which yielded radiocarbon dates between 9700 and 11,700 yr B.P. The projectile points (Fig. 6.7) are an important find, because they closely resemble the Paleoindian points found in the American Southwest. In this regard these points are unique for Eastern Beringia, because all the other projectile points found thus far that relate to late Pleistocene/early Holocene cultures in Alaska and the Yukon bear little resemblance to the Paleoindian points from the "lower 48." The oldest Paleoindian tools from sites south of late Wisconsin Ice Sheets are the Clovis tools. The Clovis complex dates from 11,200 to 10,900 yr B.P.

There are several types of fluted projectile points associated with Paleoindian sites that date to around 11,000 yr B.P., including Folsom points and others. The Mesa site points may be one of the missing links in the reconstruction of the history of Paleoindian migrations across the land bridge and south into unglaciated North America. They resemble artifacts from the Agate Basin complex (Fig. 6.7), another type of Paleoindian artifact from Wyoming described by archaeologist George Frison in 1991. Although the Mesa site points may help to answer one important question in the prehistory of North America, they also raise a few questions of their own. Probably the most important of these is: Do the Mesa site points represent a different culture than the ones archaeologists describe as the Nenana complex from the Alaskan interior? Do the Mesa site points indicate that there were two separate invasions of Alaska from Siberia during the late Pleistocene? These questions await further research and may never be fully answered.

The projectile points made by Paleoindians were well suited for hunting large mammals. Although the points themselves were not very large, they were hurled with great force. To achieve maximum velocity and distance from their projectiles, Paleoindians used the atlatl, or spear thrower (Fig. 6.8). The shaft to which the point was attached was essentially a long dart that fit into a notch on another shaft, held by the hunter. The additional length of the hand-held shaft provided added leverage in throwing, allowing the dart to travel farther and faster. This in turn allowed Paleoindians to hunt megafaunal mammals from a safer distance than if they had thrown simple spears. A wounded mammoth was undoubtedly a very dangerous animal to tangle with! The projectile points were hafted (mounted) onto dart shafts with narrow strips of wet rawhide (Fig. 6.9). When the leather

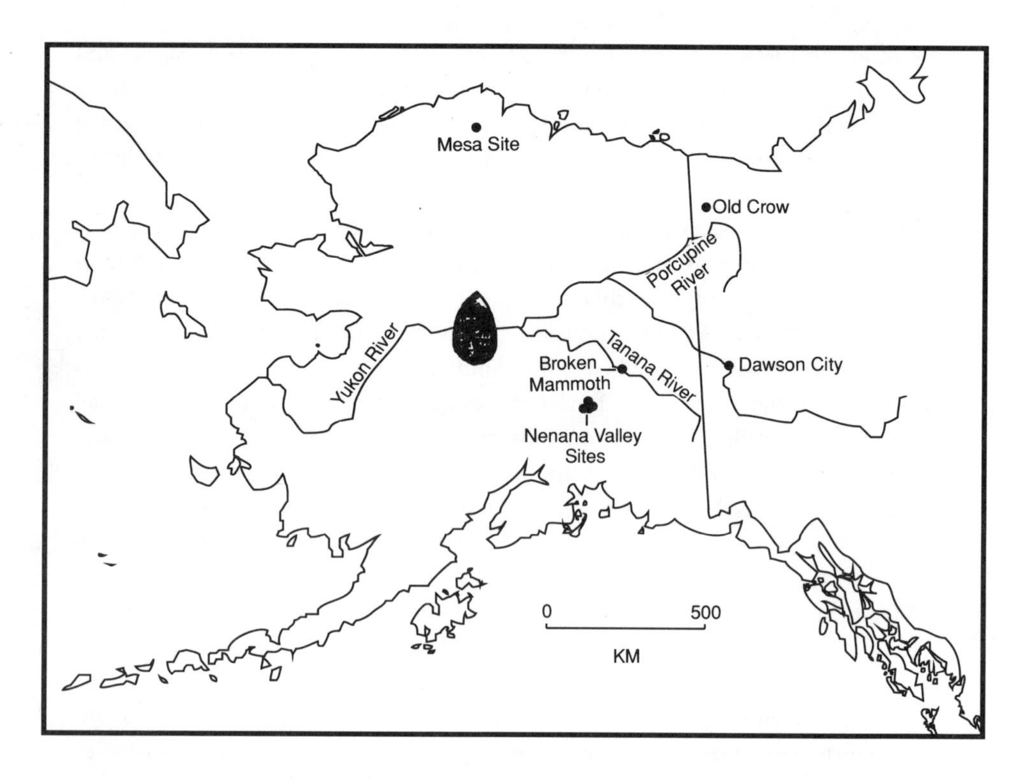

Figure 6.6. Map of eastern Siberia, Alaska, and the Yukon Territory showing locations of archaeological sites discussed in the text and a bifaced tool typical of Paleoindian sites in Alaska.

dried, it shrank to a very tight fit. Evidence from the Monte Verde site in Chile suggests that tar or pitch may also have been used to help fix a point on the end of a shaft.

Regardless of which group of prehistoric hunters made the first foray into Eastern Beringia, evidence of human settlement in Alaska and the Yukon prior to 12,000 yr B.P. remains doubtful. Human occupation of western Beringia (Siberia) began somewhat earlier. Human populations were certainly in more southern regions of eastern Asia many thousands of years earlier. In Siberia, the ancestors of the Paleoindians occupied a number of sites to the east of the Verkhoyansk mountain range (Fig. 6.10). According to John Hoffecker and colleagues, who have summarized the archaeological evidence from Beringia, this range formed a major boundary in the peopling of Siberia. West of this range, in the basin of the Lena River, archaeological sites have been found that date back to at least 20,000 yr B.P. The land east of the Verkhoyansk Range appears to have been colonized only near the end of the Pleistocene. Only two well-documented sites of late Pleistocene age have been found in this region. The Berelekh site, near the modern arctic coast, was occupied by people as early as 13,400 yr B.P. The oldest human occupation of the Ushki I site, on the Kamchatka Peninsula, has been dated at 14,300 yr B.P.

The people who came across the land bridge were Paleolithic hunters. They were nomads who followed herds of large mammals. Their tool kit consisted mostly of chipped stones, including scrapers, knives, awls, and spear or dart heads (Fig. 6.11). These tools allowed their makers to kill and butcher game animals and prepare their hides for clothing, tents, and other leather goods. With a few rare exceptions, all the leather goods have long since decomposed, and we have only a few of their stone tools, and the flakes of stone chipped off when they were made, to try to reconstruct how they lived. Archaeologists working on early human occupation sites in Alaska are of the opinion that these early hunters lived in temporary camps, moving frequently to keep up with seasonal migrations of megafaunal mammals. This life-style contrasts with what has been found in late Paleolithic sites in southern Siberia and the Russian Plain. In those regions, more or less permanent base camps were maintained. The life-style of the Eastern Beringian hunters suggests that they faced a harsher environment and were forced to keep moving in pursuit of game animals.

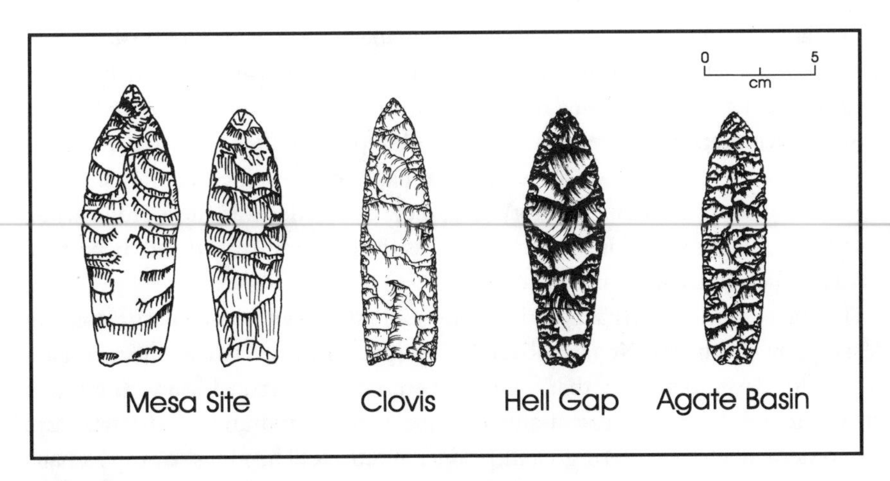

Figure 6.7. Projectile points from the Mesa site, Alaskan North Slope, compared with Clovis, Hell Gap, and Agate Basin Paleoindian projectile points from Wyoming. (The latter are from George C. Frison, *Prehistoric Hunters of the High Plains,* 2nd ed., ©1991, Academic Press, reprinted by permission.)

Figure 6.8. Drawing of a Paleolithic hunter, demonstrating the use of the atlatl, or spear thrower, while hunting a mammoth.

As far as we know, they wove no fabrics, made no pottery, built no permanent dwellings, and left no paintings or carvings on rocks or in caves. Recently, however, some nonstone tools of Beringian Paleolithic hunters have been discovered. These are caribou antlers that were shaped into **flintknapping** punch tools to drive flakes off workable stones. One of these punches has been described from a site near Dawson City in the Yukon Territory (Fig. 6.6). The antler was AMS radiocarbon dated at 11,350 yr B.P. This is nearly as old as the oldest reliably dated artifacts in Alaska, which go back as far as 12,000 yr B.P.

The biggest find of artifacts dating to the earliest known human occupations in Alaska come from the Nenana River Valley, just north of Denali National Park (Fig. 6.6). Several sites in that valley have produced **bifaced stone tools** and **projectile points** that are essentially the same as contemporaneous artifacts from the western side of the Bering Land Bridge, in Siberia. The oldest of the Nenana Valley artifacts dates to about 12,000 yr B.P. By 11,000 yr B.P., the bifaced tools were replaced by microblade tools. Microblades are small, rectangular, razor-sharp stone chips. It is thought that these blades were embedded in wooden shafts or shaped antlers for use as cutting tools.

Another early Paleoindian site has recently been discovered to the north of the Nenana Valley (Fig. 6.6). It is called the Broken Mammoth site, because small tools made from mammoth ivory were discovered there. These tools include pins and sewing needles; these are exciting discoveries because they are the first of their kind ever found in a New World Paleoindian site. The site lies on a bluff overlooking the broad floodplain of the mighty Tanana River. In addition to the usual megafaunal mammal bones discovered at some Paleoindian sites, the Broken Mammoth site also yielded the bones of geese, ducks, and other waterfowl that probably migrated up the Tanana River valley 11,500 years ago. In addition, bones and scales of salmon were found. This discovery has interested several groups of researchers. First, the archaeologists were glad to learn more about the diet and prey animals of the Paleoindians, and the paleontologists and modern zoologists were interested to learn that salmon were already making their annual migration upstream from the Yukon River to the Tanana before the end of the Pleistocene. Work continues at the Broken Mammoth site, but it has already told us a great deal that we did not previously know about late glacial life (human and otherwise) along the Tanana.

In the 1970s, some Canadian researchers startled the scientific world with discoveries of bones and antlers of mid-Wisconsin age that had apparently been shaped into tools by people. These "artifacts" were found along bluffs of the Old Crow River, in the northwest corner of the Yukon Territory (Fig. 6.6).

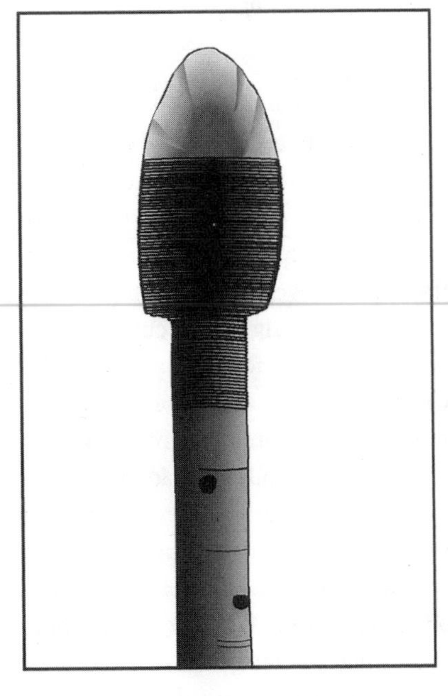

Figure 6.9. Drawing illustrating the mounting of a projectile point onto a shaft using narrow strips of rawhide.

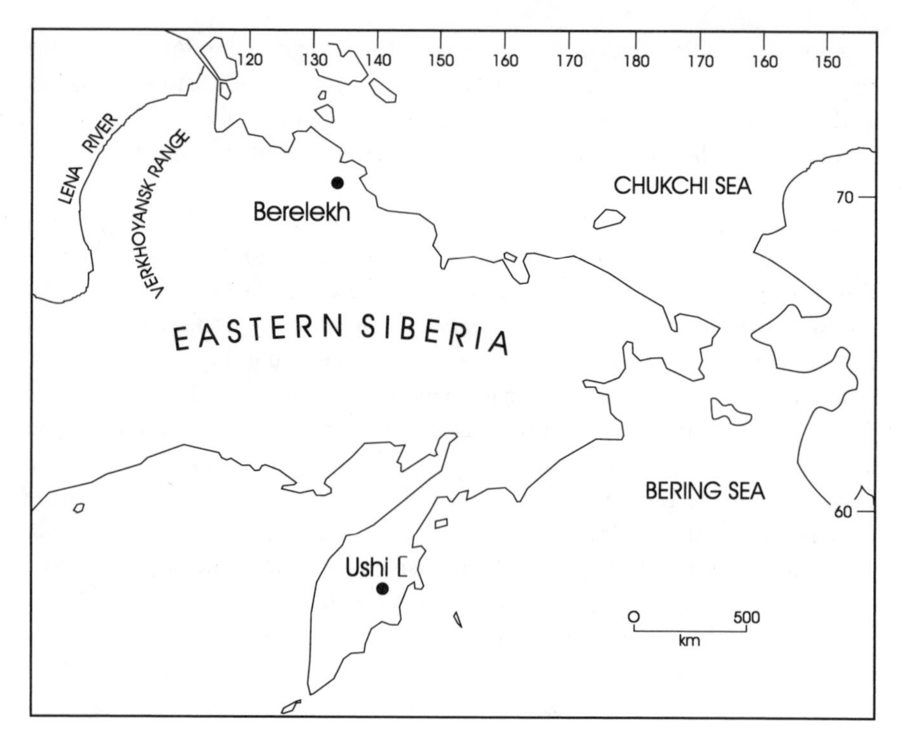

Figure 6.10. Map of eastern Siberia, showing locations of archaeological sites discussed in the text.

However, the specimens, which everyone agreed had been fashioned into tools, turned out to be much younger than originally thought. Some were only a few thousand years old. Some of the redated specimens were definitely of mid-Wisconsin age, but not all archaeologists are convinced that those specimens were really tools made by people. Some of the notable pieces were mammoth leg bones that had been broken into shapes that would be usable as digging or cutting tools.

Unfortunately, some natural phenomena can also produce fractures on bones that look very much like those on the Old Crow specimens. For instance, paleontologists have observed that fossil bones that weather out of arctic river bluffs fall down to the water's edge and are often broken to pieces through the action of river ice. The combination of freeze–thaw cycles and the scouring of riverbanks by blocks of ice moving downstream during the spring breakup of a river can produce lengthwise fractures that split bones in much the same way seen in the Old Crow mammoth specimens. Another possibility is that mid-Wisconsin mammoth bones, strewn across the banks of the Old Crow river for hundreds of years, were trampled by mammoths or other large animals and broken into the shapes now seen in the fossil specimens. This kind of trampling has been observed in dried bones of modern elephants in Africa. Finally, large carnivores or scavengers may have broken the mammoth leg bones through gnawing, in order to get at their marrow, shortly after the mammoths died.

These natural processes may have worked singly or in conjunction to produce broken bones that mimic the actions of human toolmakers. Unless or until irrefutable tools of mid-Wisconsin age are found in Alaska or the Yukon, we are left with a tantalizing enigma: were the bone "tools" really made by people? If so, why weren't any stone tools found in the same deposits or in nearby deposits of the same age? One hypothesis that explains the use of broken mammoth bones as tools is as follows: The Paleoindian hunters traveled light, carrying only the bare minimum of stone tools. When they managed to kill a mammoth, they would grab a boulder and use it to smash a leg bone into fragments. Then they would use those long, sharp bone pieces to butcher the rest of the mammoth. This method of butchering leaves no trace in the archaeological record except the broken mammoth bones themselves. A team of archaeologists who specialize in Paleoindian studies staged an experiment in which they completely butchered a dead circus elephant, using only one large rock to get things started (their only real problem was what to do with the several hundred pounds of elephant meat, when they were done).

Apparently, the Paleoindians in Eastern Beringia were a restless lot, for as soon as an ice-free route opened up (possibly about 11,800 yr B.P.), the archaeological evidence suggests that some of them moved south, following the megafauna into the rest of ice-free North America (Fig. 6.12). Soon Paleoindians in the American

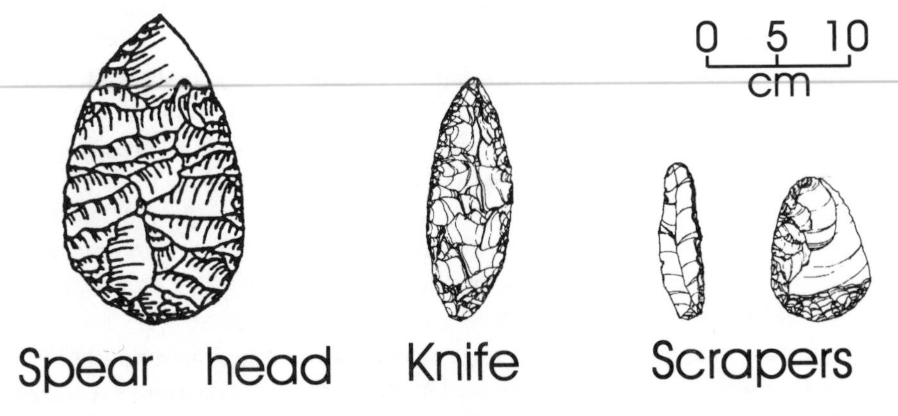

Figure 6.11. Stone tools typical of those used by the Paleolithic hunters in Beringia at the end of the Pleistocene.

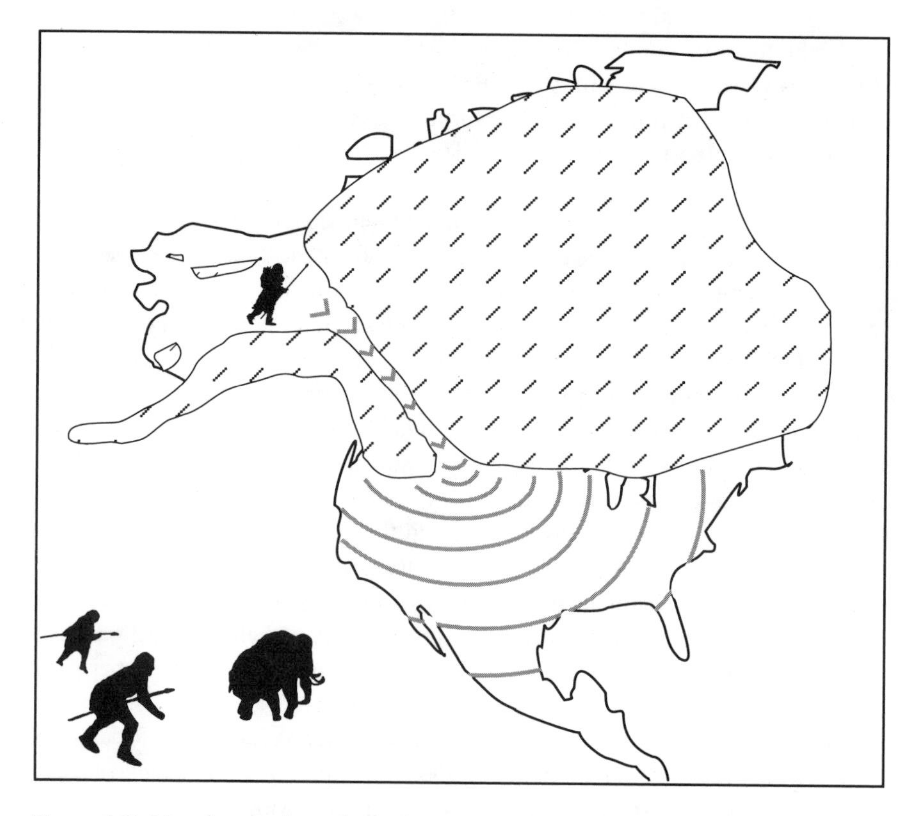

Figure 6.12. Map showing hypothesized spread of Paleoindians from Eastern Beringia to the southern ice-free regions via an ice-free corridor about 11,500 years B.P.

Southwest were fashioning the elegant bifaced projectile points known to us as Clovis points.

About 10,500 years later, they were named "Indians" by Christopher Columbus in a fit of wishful thinking. Little did Columbus know that they were actually the descendants of Siberian mammoth hunters, who had gone on to occupy all of North and South America, from subarctic Alaska to Tierra del Fuego.

The Arrival of the Inuit People

One final act was played out on the stage of arctic North America. Curiously, once the Indians were established in postglacial Alaska and Canada, they rarely ventured into the arctic regions. Probably about 6000 years ago, the first arctic-adapted people, including the ancestors of the modern Aleut peoples, began settling the western coast of Alaska and the northern coastal regions of Alaska and Canada. These people were specialists in the hunting of marine mammals, including whales, seals, walrus, and the narwhal. They made their way east from coastal Siberia by boats and settled the arctic regions. The Aleuts established themselves on the Alaskan Peninsula and the Aleutian Island chain. By 4000 years ago, the arctic hunters traveled as far east as the coasts of Labrador and Greenland. Another wave of arctic hunters, the Thule people, who specialized in hunting large whales, settled the arctic coast of North America about 1000 years ago. These are the direct ancestors of the modern-day Inuit people. When European sailors met their descendants in the arctic waters, they named them "Eskimos," adopting a version of the Cree Indian name for these people, which means "he eats it raw."

Some Alternate Sources of Information

Besides the archaeological evidence, two very different sources of information have been employed to shed light on the timing and nature of the peopling of the New World. One source comes from the field of genetics. The other comes from linguistics, the study of language. The genetic theory, proposed by Douglas Wallace, Antonio Torroni, and Theodore Schurr, suggests that all living American Indians can trace their ancestry back to one of four women, who were among the first people to migrate from Siberia to the New World. The genetic research also suggests that the first immigrants were the ancestors of people who today speak languages in the group known as Amerind. These include the Pima Indians of Arizona, the Mayans of Mexico, and the Yanomami of Venezuela. Based on genetic similarities, these groups all came from one ancestral stock, which spread out across North and South America. The genetic evidence also suggests that the first immigrants made their way into the New World much earlier than the Clovis people.

The genetic theory is based on analysis of genes from living Native Americans, ranging from Aleut and Inuit inhabitants of Alaska to the Kraho tribe in Brazil. Genetic material from mummies and preserved soft tissue in ancient skulls was also examined. The geneticists compared the genes from the part of the cell called the mitochondria, which is the energy-producing part of the cell. Mitochondrial DNA (mtDNA) is inherited exclusively from the mother, without undergoing the shuffling and mixing that occurs when the genetic material of both parents is combined at conception. Changes in mtDNA through time are easier to study than changes in other DNA, because mtDNA mutates more rapidly, so it can be used as a measuring device to establish human lineages. Also, because mutation rates are thought to be relatively stable through time, mtDNA can be used as a crude measure of time elapsed since a group's ancestors were living.

The scenario reconstructed from genetic evidence, summarized in Figure 6.13, is as follows. A band of settlers, including women carrying four distinct lineages, entered the New World sometime between 41,000 and 21,000 yr B.P. (based on mutation rates in mtDNA). The settlers found no other people to compete for space or food, so they prospered in their new environment, and their descendants quickly spread across North and then South America. These were the ancestors of the Amerind-speaking tribes. A second wave of people entered the New World, sometime between 12,500 and 6000 yr B.P. These were the ancestors of the Na-Déné speakers (the Dogrib, Tlingit, and other Pacific Northwestern tribes as well as the Athapaskan, Navajo, and Apache tribes), who are all descended from one maternal lineage, according to the genetic evidence. The ancestors of the Aleut and Inuit peoples arrived at about the same time as the Na-Déné ancestors.

The genetic research is quite controversial, and the methods employed are in need of refinement. Even though this line of evidence is only beginning to have an impact on the field of Paleoindian anthropology, it offers a fascinating, independent perspective on the subject, based on secrets hidden in the genes of the modern descendants of those early hunters.

The linguistic approach to the reconstruction of the history of American Indians also relies on modern data, namely, the languages spoken by the tribes in North and South America. Linguist Richard Rogers and his colleagues, paleontologists Larry Martin and Dale Nicklas, theorize that the current geographic location of the North America tribes can best be explained in terms of barriers to human migration that no longer exist. Looking back through time to find these barriers, Rogers and his colleagues decided that these barriers most likely were the continental ice sheets of the Wisconsin Glaciation. Human languages, like human genes, are not static. They undergo changes through time. (Indeed, how many of us can understand the English language as it was spoken a thousand years ago?) Using estimates of rates of language change through time, linguists have extrapolated how long one group of speakers (i.e., one tribe) has been isolated from another group when both groups' languages are derived from one common ancestral tongue.

The linguistic theory, like the genetic theory, places Paleoindians in North America long before the earliest Clovis occupation. Rogers and his colleagues believe that the ancestors of modern American Indians arrived some time during the last glaciation and that events at the end of the ice age were an important force in shaping the modern distribution of tribal language groups. The linguists speculate that Paleoindians lived in central and eastern North America, in regions south of the Laurentide Ice Sheet, during the Wisconsin Glaciation. Then, when the ice began to retreat, four major language groups (the Algonquian, Iroquoian, Siouan, and Caddoan) began spreading north (Fig. 6.14).

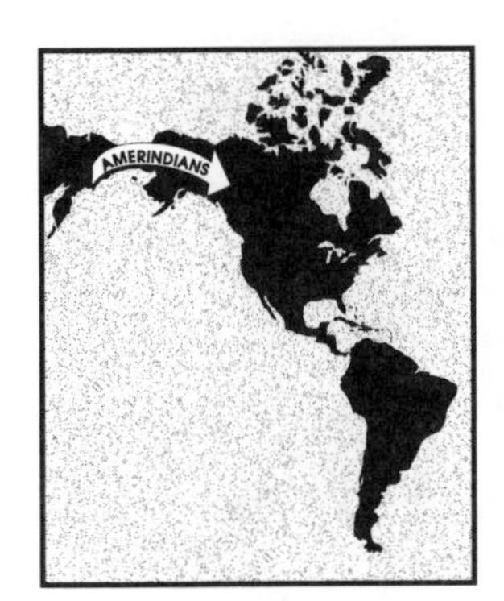

FIRST WAVE:
Archaeological Evidence: ca. 12,000 yr B.P.
Genetic Evidence: 42,000-21,000 yr B.P.
Linguistic Evidence: before 12,000 yr B.P.

SECOND WAVE:
Archaeological Evidence: ?
Genetic Evidence: 12,500-6000 yr B.P.
Linguistic Evidence: pre-12,000 yr B.P.

THIRD WAVE:
Archaeological Evidence: ca. 5000 yr B.P.
Genetic Evidence: 12,500-6000 yr B.P.
Linguistic Evidence: after 11,000 yr B.P.

Figure 6.13. Maps showing summaries of proposed theories on the timing of arrival of three principal groups of Native Americans into the New World, as discussed in the text.

Figure 6.14. Map showing hypothesized movements of major language groups of Native Americans, following the retreat of Wisconsin Glaciation Ice Sheets.

These changes and migrations led to a redistribution of language groups. The Caddoan, Siouan, and Iroquoian speakers took advantage of the northward spread of the broad-leafed deciduous forest zone. The Algonquian speakers followed the receding Laurentide ice margin north into the eastern Canadian subarctic zone. Further west, Athapaskan speakers (part of the Na-Déné group) moved inland, occupying land uncovered by the retreating Cordilleran Ice Sheet. Eskimo-Aleut speakers are thought to have survived the Wisconsin Glaciation along the coasts of the Bering Land Bridge. They used their maritime hunting skills to take advantage of the newly exposed arctic coastal region of Canada, spreading east across the arctic region during the last 6000 years.

This theory, like the genetic theory, is interesting because it is totally independent of the archaeological evidence of Paleoindians, which is composed almost exclusively of stone tools. As such, it is tantalizing, if unprovable. Again, if more refinements are made, this linguistic approach may make a useful adjunct to the physical science approaches. Possibly these theories will all work in concert to improve our understanding of the peopling of the New World. Be that as it may, one of the most commonly agreed-upon aspects of this fascinating story is that it all began when a band of ancient hunters hiked up a hill, where they left the broad plains of the Bering Land Bridge and entered the New World in the region we call Alaska.

Suggested Reading

Bryan, A. L. (ed.). 1986. *New Evidence for the Pleistocene Peopling of the Americas.* Orono, Maine: Center for the Study of Early Man. 368 pp.

Dixon, E. J. 1993. *Quest for the Origins of the First Americans.* Albuquerque, New Mexico: University of New Mexico Press. 154 pp.

Frison, G. C. 1991. *Prehistoric Hunters of the High Plains,* second edition. New York: Academic Press. 532 pp.

Gibbons, A. 1993. Geneticists trace the DNA trail of the first Americans. *Science* 259:312–313.

McGhee, R. 1978. *Canadian Arctic Prehistory.* Toronto: Van Nostrand Reinhold. 128 pp.

Powers, W. R., and Hoffecker, J. F. 1987. Late Pleistocene settlement in the Nenana Valley, central Alaska. *American Antiquity* 54:263–287.

Rogers, R. A., Martin, L. D., and Nicklas, T. D. 1990. Ice-age geography and the distribution of native North American languages. *Journal of Biogeography* 17:131–143.

West, F. H. 1983. The antiquity of man in America. In Porter, S. C. (ed.), *Late-Quaternary Environments of the United States.* Minneapolis: University of Minnesota Press, pp. 364–382.

7

KENAI FJORDS NATIONAL PARK
An Ever-changing Landscape

Fjord is a Norwegian word for a narrow inlet of the sea between two steep slopes or cliffs. (The American spelling is more commonly *fiord,* but I will use the original spelling here.) Norway is famous for its fjords, but equally spectacular scenery can be found in Alaska, at Kenai Fjords National Park. The park contains some magnificent glacially deepened fjords on its rugged coast. The glaciers at the ends of the fjord inlets descending from the mountains and the Harding and Grewingk–Yalik Icefields dominate the upland landscapes. The Harding Icefield is the largest one that is totally within the United States. The dominant physical features of the park are the three major fjord systems: Aialik, Northwestern, and McCarty. These contain numerous fjord-calving tidewater glaciers. Bear Glacier, adjacent to Resurrection Bay, is the second longest outlet of the Harding Icefield. Yalik and Petrov glaciers are among those that drain from the smaller Grewingk–Yalik Icefield. The physical environment of this region is dominated by glacial geology, but in spite of all the glacial activity (past and present), Kenai Fjords is far from a sterile landscape of rock and ice. The region also has a fascinating and unique ecological history.

Modern Setting

The Kenai Peninsula is located near the edge of the modern limit of **Pacific coastal forest,** which extends more than 4000 km (2500 miles) in an arc from northern California to Kodiak Island, Alaska. This type of forest grows where summers are cool, winters are not too cold, and precipitation is high. At the northern end of the coastal forest region, the trees give way to tundra. Inland in Alaska, the luxuriant coastal forest is replaced by spruce–birch forests, adapted to drier, more continental climates where summers are warmer and winters are much colder. The dominant trees in the coastal forests of the Kenai Peninsula are Sitka spruce and mountain hemlock, with fewer stands of western hemlock and Alaskan yellow cedar than are found in forests farther east on the Alaskan Gulf coast.

Sitka spruce dominates the upper forest limit on mountain slopes. Treeline ranges from about 300 to 600 m (1000 to 2000 ft) elevation in the Kenai Mountains, depending on slope and aspect. Alpine tundra regions are covered by sedges, grasses, and heaths, with dwarf shrub communities of willow, birch, and alder growing in moist patches.

The modern climate of the Kenai Peninsula is predominantly cool and wet. Average annual precipitation ranges from 600 mm (23 in.) at some high elevations to 1700 mm (66 in) at sea level. The Kenai Mountains form a rainshadow, with the lee (western) side receiving less precipitation than coastal regions of the Gulf of Alaska. The landward decrease in precipitation is accompanied by a corresponding decrease in mean annual temperature. This phenomenon holds true throughout Cook Inlet, and it is reflected by the modern glaciers of this region. Large glaciers on slopes facing the Gulf of Alaska feed from ice fields ranging from 900 to 1800 m (2950 to 5900 ft) elevation. Twenty kilometers (12 miles) inland, small cirque glaciers survive only on north-facing slopes, at elevations over 1250 m (4100 ft).

The varied terrain of the Kenai Peninsula, including mountains, forests, rolling hills, wetlands, rivers, streams, and lakes, supports an abundant and diverse fauna, including marine mammals in coastal waters. Moose are particularly plentiful. This may be the result of several large fires in recent decades. The fires opened up tracts of forest to the invasion of birch and aspen, favored browse vegetation for moose. About 25 to 30 pairs of trumpeter swans can be found nesting in shallow lakes of Kenai Fjords National Park in a given year. The park's highlands are home to Dall sheep, mountain goats, caribou, and ptarmigan. Lower mountain slopes are inhabited by brown and black bears, wolves, wolverines, beavers, loons, grouse, and bald eagles that fish for salmon along streams and rivers.

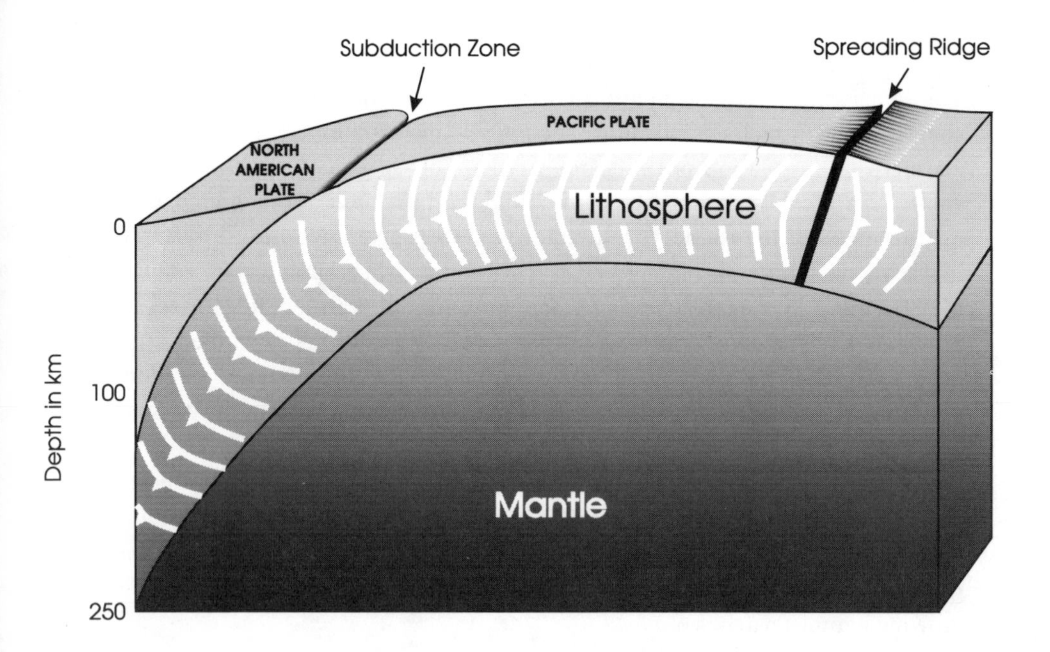

Figure 7.1. Theory of tectonic movement of an oceanic plate toward a continental plate along the Gulf Coast of Alaska. As the slab descends, it pulls the crust, or lithosphere, along behind it, causing an opening of an oceanic ridge.

Geologic History

To understand the geography of Kenai Fjords National Park, it is necessary to become familiar with the theory of **plate tectonics.** The Earth's crust, and the solid upper part of the dense mantle rocks below, are composed of giant slabs called plates, which currently move between 2 and 10 cm (1–4 in.) per year. As far as geologists can tell, these plates have never sat still for very long. In fact, both the geologic and biological history of our planet have largely been shaped by these movements. Ridges, trenches, and rifts outline the plates, and some of the most dramatic tectonic effects take place along these boundaries between plates. The plates leave huge cracks at the ridges where they spread and create subduction zones where one plate is forced under another. The tremendous forces and friction along these subduction zones are sufficient to melt large quantities of rock. Because the crust has many cracks along these zones, the molten rock finds its way to the surface in the form of volcanoes. The Gulf coast of Alaska sits astride one of these subduction zones, where the Pacific plate (an oceanic plate) meets the North American plate (a continental plate) (Fig. 7.1). Geologists estimate that the Pacific

Ocean and North American Continental plates have been converging in this region since the Triassic Period, roughly 200 million years ago.

This subduction has created many of the geographic features of southern Alaska, including a deep trench that runs along the margin of the Aleutian Island chain, a zone of increased seismic activity (an earthquake zone) along the coast of south-central Alaska, and a chain of volcanoes along the Alaskan Peninsula, some of which formed during the Quaternary Period. The Kenai Peninsula lies along the subduction zone, between the volcanoes of the Alaskan Peninsula and Aleutian Islands and the Aleutian trench.

During the late Quaternary, the Kenai Peninsula has subsided more than 100 m (300 ft). The most recent episode of large-scale subsidence occurred during the 1964 Alaska earthquake, when the Kenai region sank more than 1.2 m (4 ft) in a day. Drowned forests can be seen at several localities along the peninsula. These stands of trees sank below sea level during the earthquake and were killed by the salt water.

Pleistocene Glaciations

In addition to plate tectonics, other geologic forces have been at work in shaping the Kenai region that we see today. Chief among these are the glaciations that took place during the Pleistocene. Major icefields persist today in the highland regions of Kenai Peninsula (Fig. 7.2), but these are far smaller than their predecessors in the Pleistocene.

During the last (Wisconsin) glaciation, all but the highest peaks on the Kenai Peninsula were buried by ice. This was only the last chapter in the glacial story, however. Glacial geologist Thor Karlstrom (1964) described the Kenai landscape as made up of a "bewildering array of landforms, most of which reflect the complex history of multiple glaciations." Quaternary sediments mantle the Kenai lowlands in thicknesses of up to 230 m (750 ft). The large lakes, Skilak and Tustumena, fill glacially scoured basins rimmed by glacial moraines. Large moraines record at least two and perhaps three major glacial events, in which ice flowed eastward from the Alaska–Aleutian Range and westward from the Kenai Mountains.

The history of Pleistocene glaciations on the Kenai Peninsula remains somewhat unclear, but geologists Henry Schmoll and Lynn Yehle of the U.S. Geological Survey have proposed the following scenario about the timing and extent of some of the more recent glaciations. The most extensive Pleistocene glaciation in this region, the Eklutna Glaciation, began about 200,000 yr B.P. Glacial ice covered regional mountain ranges and extended out great distances beyond the modern coastline, filling the straits between the Kenai Peninsula and Kodiak Island. Glaciers filled the Susitna, Matanuska, and Turnagain Arm valleys, coalescing with ice

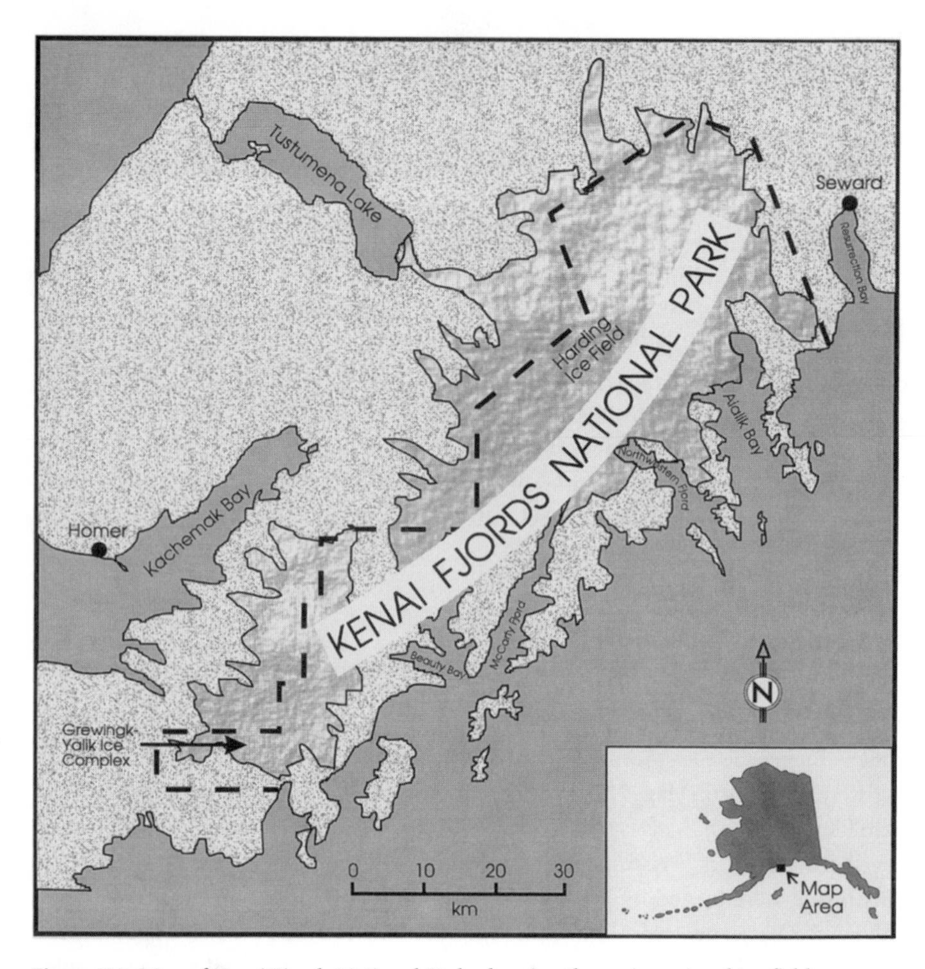

Figure 7.2. Map of Kenai Fjords National Park, showing the main regional ice fields.

spilling off the flanks of the Alaska Range and Kenai Mountains. Ice all but covered the Caribou Hills except for regions above 600 m (2000 ft) elevation. These regions served as **nunataks:** regions above the ice that may have supported some plant and animal communities throughout the glaciation. There is some circumstantial evidence for this theory in the modern flora of the region, which contains some species of **endemic** alpine plants that are thought to have survived the last glaciation in regional alpine localities.

The Knik Glaciation (circa 75,000–53,000 yr B.P.) was not as extensive as previous glaciations. Glaciers spilled into the Cook Inlet basin but did not fill it. Major ice lobes flowing through the Kenai, Skilak, Killey, and Tustumena Valleys from the Harding Icefield coalesced along the margins of the Kenai Lowland to form a continuous ice shelf, 40 km (25 miles) wide and 80 km (50 miles) long, parallel to the Kenai Mountain front north of the Caribou Hills. Karlstrom estimated that the ice was 450 to 600 m (1500 to 2000 ft) thick near the head of Tustumena Lake.

The Naptowne Glaciation, the last of the Pleistocene glaciations, spanned the interval of 30,000–12,500 yr B.P. Recent work on the Kenai Peninsula suggests that Naptowne-age advances were extensive, once again covering all but the upper Caribou Hills, between the town of Homer and Lake Tustumena. The magnitude of the forces involved in the movement of glacial ice can be seen in the glacial outwash deposit laid down as the Tustumena Glacier receded (Fig. 7.3) and in the huge **glacial erratics** (Fig. 7.4). These are boulders that were gouged out of bedrock by glacial ice, then carried along with the ice flow, and eventually dropped as the ice receded.

Geologic evidence suggests that during the late Wisconsin interval, meltwater from glaciers north of Anchorage was dammed by late-lying ice lobes in Cook Inlet, forming a **proglacial lake** (Fig. 7.5). This has been named Glacial Lake Cook. Lake levels rose and fell repeatedly as meltwater breached the ice dam at the spillway. Such glacially dammed lakes have also formed in recent times in southern Alaska; they are inherently unstable and given to catastrophic drainage. The deglaciated shores of Glacial Lake Cook may have provided some of the first ice-free habitat for regional biota in the late Wisconsin interval.

The Role of Glaciers in Fjord Formation

During the Pleistocene, glacial ice repeatedly filled the drainages leading from the Kenai Mountains to the sea. The scouring action of glaciers is thought to be the most important element in the creation of the fjords, which are narrow glacial valleys partially filled by the sea. Fjords flank many coasts at high latitudes, from Norway and Greenland across arctic Canada to Alaska. The deepening of a stream valley by glacial ice is illustrated in Figure 7.6. Glaciers gradually act to deepen a relatively shallow valley (Fig. 7.6A); during multiple episodes of glacial advance (B) and retreat (C), the valley becomes deepened. This deepening eventually brings the elevation of the valley bottom below sea level. When sea level rises during interglacials, and the glacier is in a recessional stage, water inundates the lower reaches of the valley, forming a fjord (D). However, this sequence of events is not always straightforward, because advances and retreats of tidewater glaciers (glaciers terminating in the sea) are not always synchronous with climatic cooling and warming.

Many fjords have very deep channels, but terminal moraines left by Late Wisconsin glaciers during stands at their outer limits sometimes form sills across

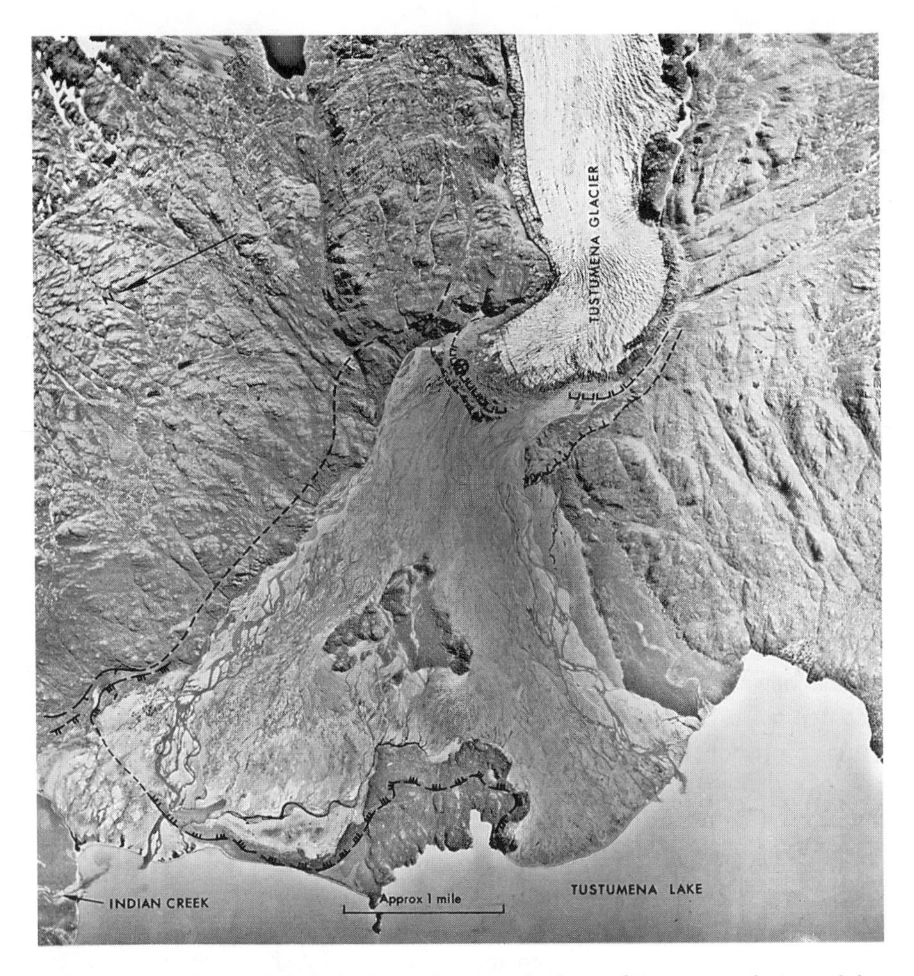

Figure 7.3. Moraines and outwash deposit between the front of Tustumena glacier and the head of Tustumena Lake. These deposits were thought by Karlstrom to have been late Holocene in age but have subsequently been radiocarbon dated to the late Wisconsin (Naptowne) glaciation. (Photograph by T. N. V. Karlstrom, U.S. Geological Survey.)

widened portions of the fjord. Typical of these is Northwestern Fjord, which David Miller reports to be 234 m deep (768 ft) in midchannel but is navigable only by kayak or other shallow-draft boats because of a treacherous spit and offshore rocks, representing glacial debris from a terminal moraine (Fig. 7.7). The pre-Wisconsin fjord limits, which are often far beyond these moraines, are now largely submerged. Many cirque basins, cut far above sea level at these outer fjord margins, are now partly or wholly submerged. These basins, once at the heads of glaciers, are now providing safe havens for small boats in rough seas.

Figure 7.4. Photograph of glacial erratic boulder near Boulder Point, Kenai Peninsula. This boulder represents ice-rafted material dropped during the retreat of Naptowne ice from Boulder Point. (Photograph by T. N. V. Karlstrom, U.S. Geological Survey.)

The fjords comprise twelve major embayments along the Gulf coast of Kenai Fjords National Park. The heads of the fjords are fed by more than 40 glaciers, many of which remain unnamed. The largest ice body in the park, Harding Icefield, is fed by more than 20 m of snowfall per year.

Holocene Glacial Dynamics

Glacial geologist Parker Calkin and associates have been studying the waxing and waning of glaciers on Kenai Peninsula that have taken place in the Holocene. Unlike other parts of the world, the Gulf coast of Alaska is cool and wet enough to support large mountain glaciers, even throughout times of interglacial warming. Wiles and Calkin studied Holocene changes in glaciers terminating in McCarty Fjord in Kenai Fjords National Park. McCarty Glacier is a **tidewater glacier,** ending in seawater rather than on land. Tidewater glaciers spawn icebergs into the water. Such glaciers are common along the Alaskan Gulf coast. By studying air photographs and the rings of trees that have been killed by glacial advances along

McCarty Fjord, these geologists were able to piece together the recent history of this glacier–fjord system. The evidence suggests that the main trunk of McCarty Glacier advanced at about 3600 yr B.P. About A.D. 550, the trunk glacier again advanced after an undetermined interval of retreat. This later readvance occurred at the same time as advances by some land-terminating glaciers on the Kenai Peninsula. Another advance by McCarty Glacier occurred at A.D. 900 during an interval generally considered to have been warm (the Medieval warm period). This was a time when most land-terminating glaciers were receding.

During the twentieth century, the glacial ice tongue retreated from McCarty Fjord at a surprising rate. From a maximum extent recorded in 1905, the ice margin retreated 20 km up the fjord in just 55 years (Fig. 7.8). During the 1940s, the glacier was retreating at a rate of 4.6 km per year. This type of rapid retreat is facilitated by iceberg calving into the sea, whereas retreat of land-terminating glaciers is generally much slower because it involves a gradual thinning of the ice body rather than the release of large blocks of ice into the water. I will discuss this phenomenon more in the next chapter. The important point to keep in mind is that tidewater glaciers represent extremely dynamic systems, and their advances

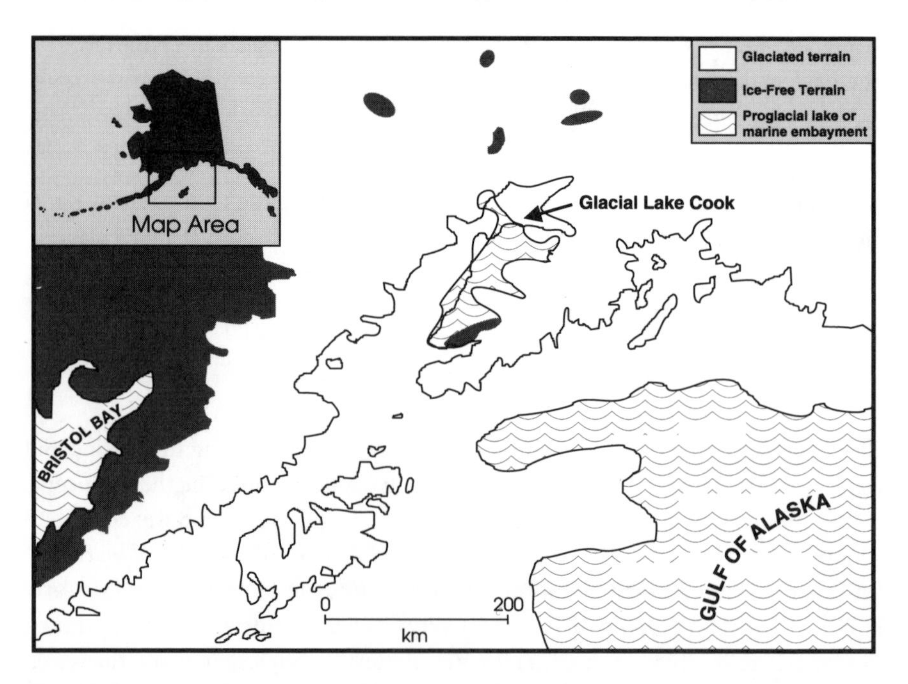

Figure 7.5. Proposed boundaries of glacial ice and Glacial Lake Cook during the late Wisconsin glaciation. (After Hamilton and Thorson, 1983.)

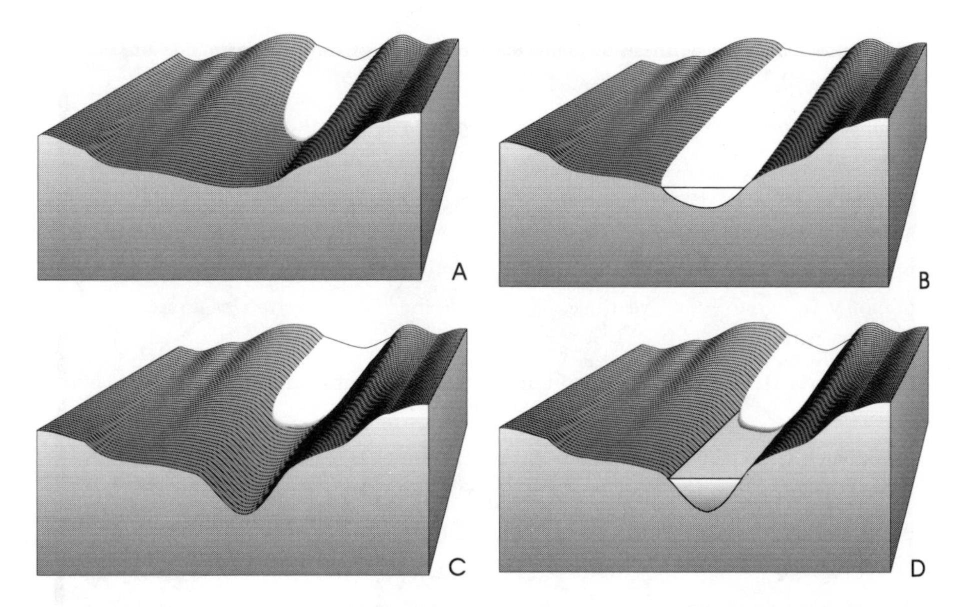

Figure 7.6. Series of cartoons illustrating the development of a fjord by glacial scouring of a drainage basin. (A) A glacier expands down a shallow stream valley. (B) Glacial ice fills the valley, cutting into the valley walls to form a U-shape. (C) The glacier retreats, leaving overdeepened valley. (D) As sea level rises during deglaciation, the sea drowns the front of the scoured basin, forming a fjord.

and retreats may or may not be in synchrony with global or regional climate changes.

Development of Postglacial Ecosystems

Sediment cores from lakes on the Kenai Lowland register the return of plant life to the region after 14,500 yr B.P. (Fig. 7.9). This date is quite early compared with dates from the oldest postglacial sediments from sites on nearby Kodiak Island and at Icy Cape, east of Anchorage (Figs. 7.9 and 7.10). The earliest plant communities described from these two regions are dated at 9500 and 10,800 yr B.P., respectively. Further east on the Alaskan Gulf coast, vegetation became established by 14,000 yr B.P. The Prince William Sound region was still dominated by glaciers coming out of the Chugach and Kenai Mountains at this time, whereas adjacent regions were becoming free of ice.

Tom Ager's Hidden Lake pollen record indicates that the earliest vegetation to become established after deglaciation was herbaceous tundra, dominated by

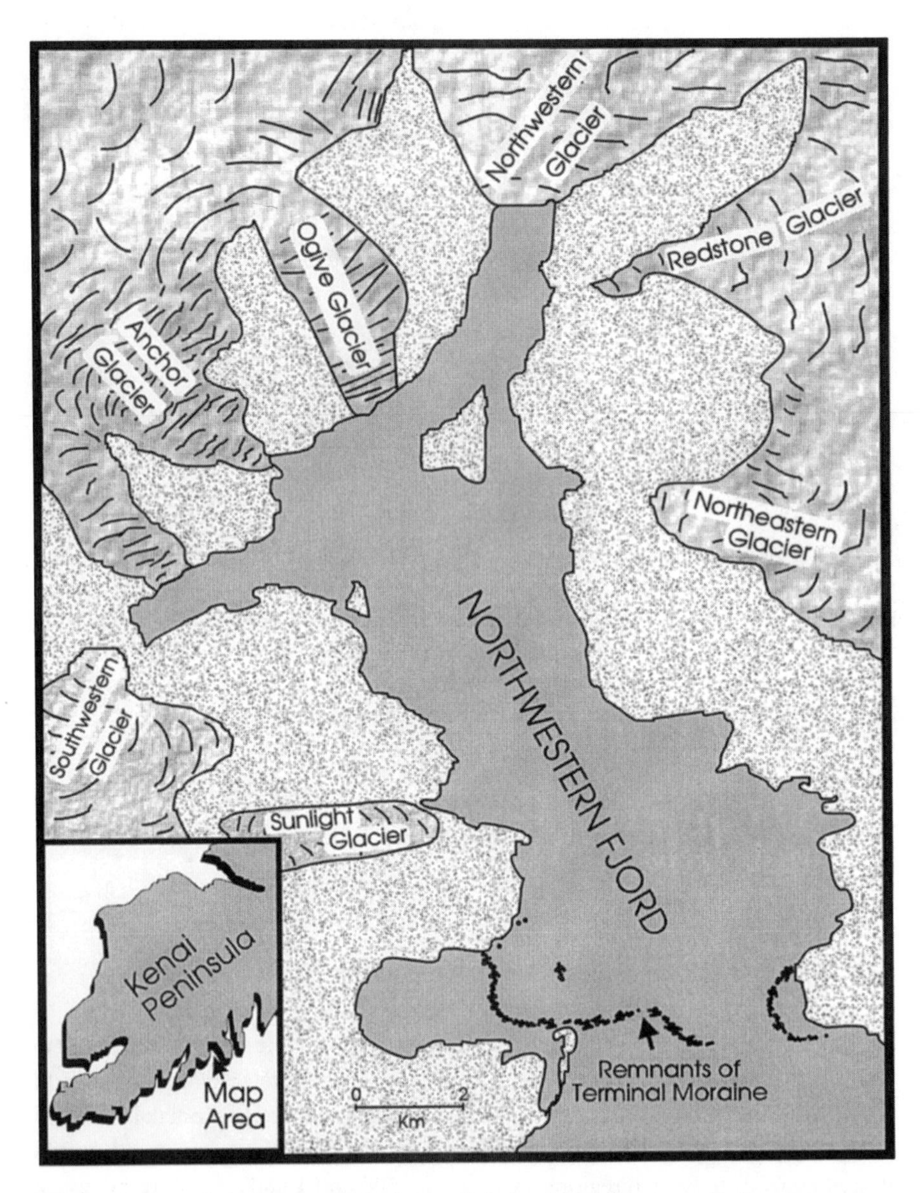

Figure 7.7. Map of Northwestern Fjord, Kenai National Park. Note terminal moraine debris blocking most of the fjord inlet. (Modified from Miller, 1987.)

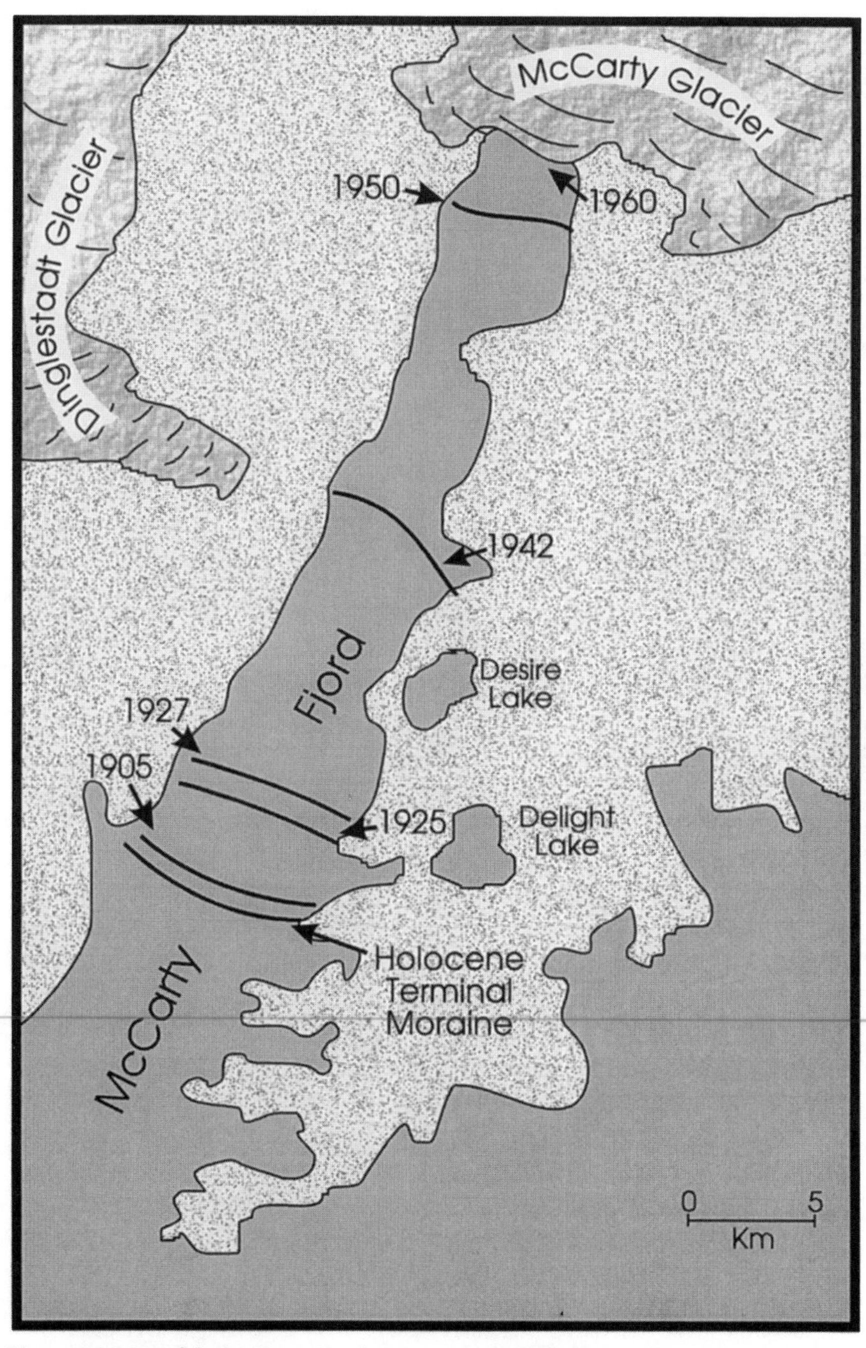

Figure 7.8. Map of the McCarty Fjord region, Kenai Fjords National Park, showing position of ice margins since 1905. (After Wiles and Calkin, 1993.)

sedges, grasses, sage, and plants in the composite family. Ager interpreted this early postglacial vegetation as having been a mosaic of plant communities growing in patches and not completely covering the deglaciated landscape. This herbaceous tundra, not unlike the steppe-tundra plant communities recorded from the Alaskan interior, was apparently short-lived on the Kenai Peninsula. In addition, by the time postglacial vegetation began colonizing adjacent regions (i.e., Kodiak Island and Prince William Sound), this type of herbaceous tundra played no part in the succession of plant communities (Fig. 7.9).

By 13,700 yr B.P. at Hidden Lake, herbaceous tundra gave way to shrub tundra, dominated by dwarf birch and heath plants. The dramatic expansion of shrub

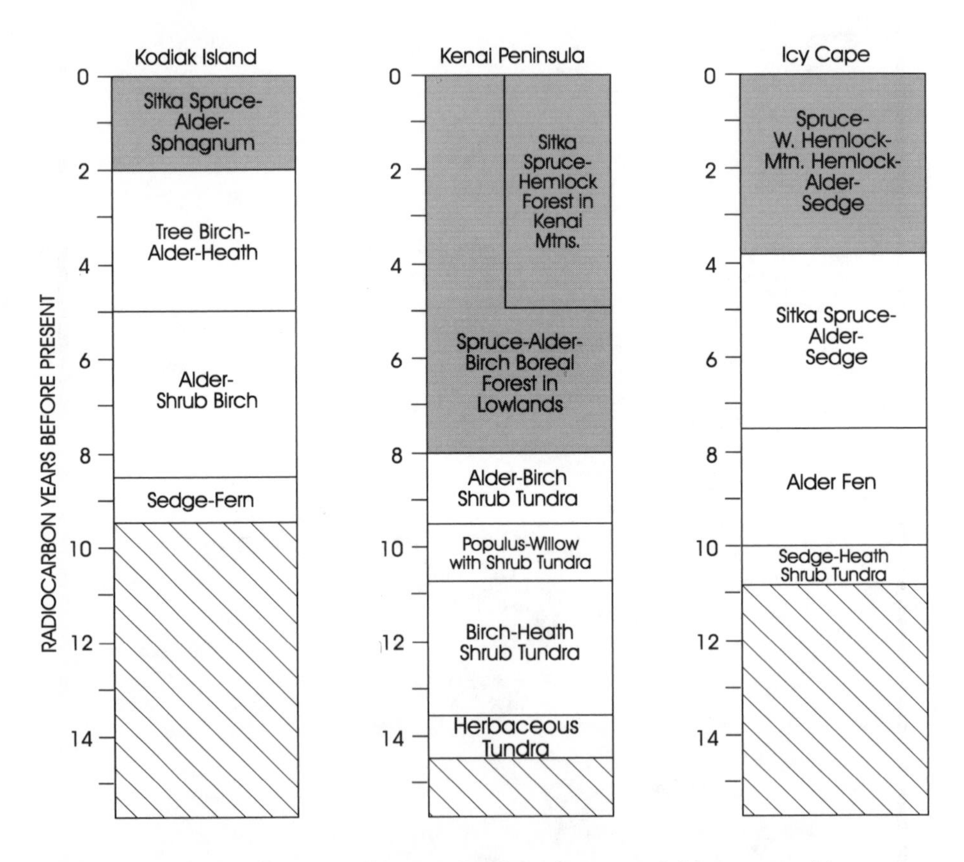

Figure 7.9. Summary diagram of the postglacial development of plant communities at three localities on the Gulf of Alaska. Kodiak Island and Icy Cape data from Heusser (1985). Kenai Peninsula (Hidden Lake site) data from Ager (1983). Site localities are shown in Figure 7.10.

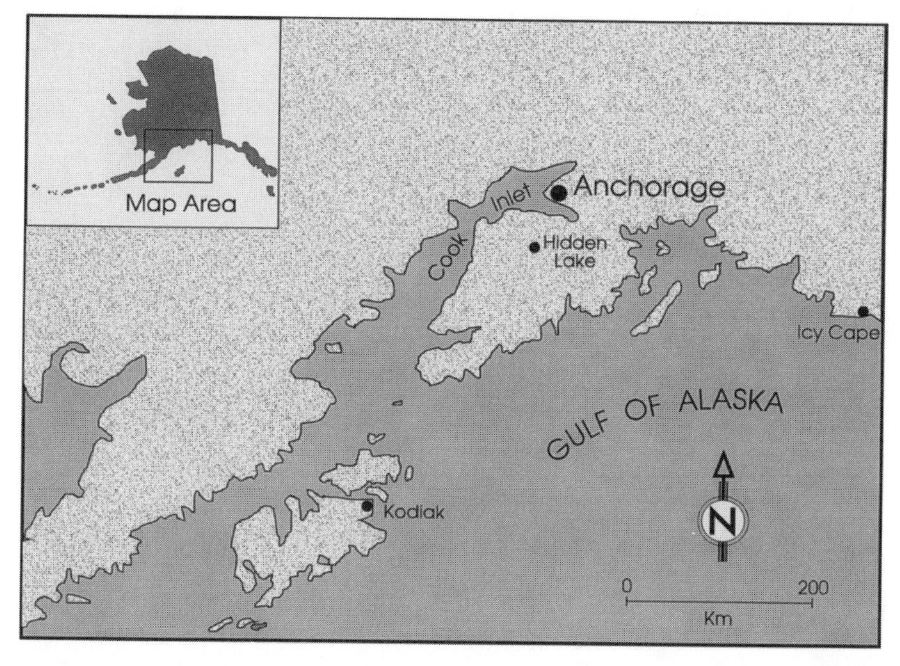

Figure 7.10. Map of the western Gulf of Alaska region, showing fossil sites discussed in the text.

birch on the Kenai Peninsula, as elsewhere in Alaska, was probably brought about by rapid climatic warming.

Elements of deciduous forest were established on Kenai Peninsula about 10,300 yr B.P., as dwarf birch shrub tundra gave way to a mixture of shrub tundra and deciduous scrub forest in which *Populus* (cottonwood, balsam poplar, and aspen) and willow were important. Alder began invading the region by about 9500 yr B.P., and within 500 years, it was a dominant species in forests throughout the Cook Inlet region.

Conifer trees began arriving on the central Kenai Peninsula about 8000 yr B.P. The first kind of conifer to be established was spruce, probably both white spruce and black spruce. These species apparently spread outward from interior Alaska early in the Holocene. Mountain hemlock and western hemlock became established between about 5000 and 4000 yr B.P. Western hemlock grows in the Cook Inlet region but is not common. Pollen records from the east coast of the Kenai Peninsula indicate that Sitka spruce and mountain hemlock may have arrived simultaneously in mid-Holocene times. Coastal forest trees, including Sitka spruce and the two hemlock species, apparently did not reach the western side of the Kenai Peninsula until the mid- to late Holocene.

It is instructive to compare the Hidden Lake pollen record with those studied from adjacent regions in Alaska. For instance, the history of postglacial vegetation on nearby Kodiak Island is quite different from that found on the Kenai Peninsula. Deglaciation was well under way on Kodiak by 12,000 yr B.P., but remnants of ice apparently persisted until nearly 9000 yr B.P. Calvin Heusser described a progression of pollen zones, beginning with the earliest postglacial plant communities, after 9000 yr B.P. This pollen zone is indicative of a sedge–fern community. Some older peat deposits on the island, dated at 11,930 yr B.P., contain pollen indicative of sedge and heath, with minor elements of willow, grasses, and composites. This older plant community is similar to that growing on the Kenai Peninsula at that time, except that the Kodiak flora lacked shrub birch. Shrub birch communities, with alder, became established on Kodiak Island after 8500 yr B.P. About 5000 yr B.P., tree birch, alder, and heath vegetation were dominant. Finally, Sitka spruce invaded the island during the last thousand years. Sphagnum bogs have also become more prevalent in the late Holocene.

Heusser also described the succession of plant communities following regional ice retreat at Icy Cape, a site farther east on the Gulf of Alaska coast (Fig. 7.10). The postglacial period also began later at Icy Cape than at Hidden Lake on the Kenai Peninsula. The earliest vegetation record at Icy Cape is dated at 10,800 yr B.P. This was a shrub tundra dominated by sedge and heath. By 10,000 yr B.P., this pioneer vegetation was invaded by alder. Sitka spruce arrived about 7500 yr B.P., and western hemlock became established after 3800 yr B.P., followed by mountain hemlock somewhat later. In recent times, alder has declined in the Icy Cape region, and closed conifer forest is now dominant throughout Gulf coast regions to the east.

Perhaps the most striking feature of the vegetation history of these regions is that the plant communities are constantly shifting. Even now, 10,000 years after the departure of glacial ice, new tree species are becoming dominant, and old species are becoming rarer. This aspect of vegetation dynamics is illustrated in Figure 7.9. The time period characterized by "modern" vegetation types is shaded in gray. That portion of the vegetation record is relatively recent, compared to many other regions of Alaska. The other interesting phenomenon observable in the figure is the length of time between the establishment of the oldest pioneer vegetation and the arrival of "modern" vegetation. For all three regions, the succession from pioneer tundra vegetation to Pacific coastal forest took many thousands of years. On the Kenai Peninsula, this progression took about 6500 years; at Icy Cape, it took about 7000 years; and on Kodiak Island, it took more than 8500 years. One might expect the development of "modern" forest to take longest on Kodiak because the spread of trees is slowed by a saltwater barrier of about 35 km.

Heusser traced the postglacial migration of Pacific coastal forest from its late Wisconsin refuge south of the Cordilleran Ice Sheet in Washington to its eventual colonization of Kodiak Island only 1000 years ago (Fig. 7.11). The Pacific coastal forest has been documented from 30,000-year-old fossil records in west-central Washington and locations farther south.

The **Cordilleran Ice Sheet** reached its late Wisconsin maximum between about 15,000 and 14,000 yr B.P.; then it began to retreat. The coastal forest moved rapidly into ice-free regions of southwestern British Columbia, becoming established by 12,500 yr B.P. Spruce and hemlock trees spread northward, reaching the southern tip of the Alaska Panhandle by about 10,000 yr B.P. Although the northern parts of the panhandle were deglaciated by about 10,300 yr B.P., the important tree species of the Pacific coastal forest did not spread as far north as this until about 7500 yr B.P.

For reasons that remain unclear, the Pacific coastal forest spread very slowly along the Gulf coast of Alaska during the Holocene. As Figure 7.11 indicates, the regions studied thus far were free of ice for as much as 8500 to 9000 years before the coastal forest ecosystem became completely established. The spread of forest was not quite as slow on Kenai Peninsula, where the lowlands were ice-free by 14,500 yr B.P. and modern forest was established by 8000 yr B.P. Even on the Kenai Peninsula, however, the hemlock element of the coastal forest apparently did not arrive on the mountain slopes facing the gulf until about 5000 yr B.P.

In most cases, the two species of hemlock (mountain hemlock and western hemlock) appear to have been the last elements of the coastal forest to arrive in Alaska. These trees generally became established after the arrival of Sitka spruce, the other important tree species in this type of forest. Sitka spruce arrived at Icy Cape by 7500 yr B.P., and the two hemlock species became established about 3700 years later (Fig. 7.9).

Could there be differences in the climatic tolerances between the spruce and the hemlocks that kept the hemlocks out of the region until the late Holocene? The differences recorded by foresters are subtle but perhaps significant. All three species can survive roughly the same degree of winter cold (–35 to –45°C, or –31 to –49°F) and summer warmth (18 to 21°C, or 64 to 70°F). The minimum number of frost-free days required for these trees ranges from 40 days for mountain hemlock to 50 days for Sitka spruce and 100 days for western hemlock. The only sharply contrasting difference in the climatic tolerances of these trees is that Sitka spruce can grow in regions with annual precipitation of from 1400 to 5000 mm of moisture, whereas mountain and western hemlock grow in regions that receive from 450 to 1800 mm of precipitation. In other words, Sitka spruce can tolerate wetter conditions than either mountain or western hemlock can tolerate. If early- and mid-Holocene climates were ex-

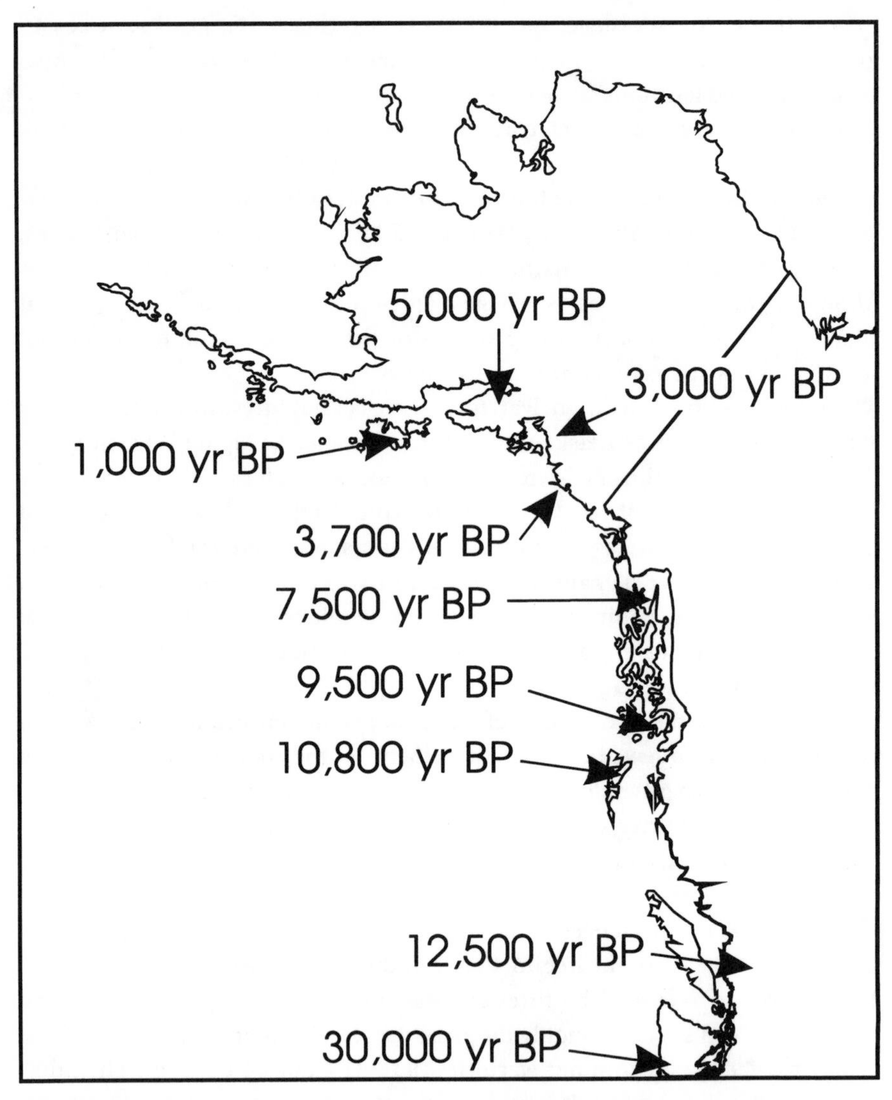

Figure 7.11. History of the postglacial spread of Pacific coastal forest from Washington to Alaska, showing radiocarbon age of first known establishment of spruce–hemlock forest. (Data from Ager, 1983, and Heusser, 1985.)

tremely wet along the Alaskan Gulf coast, then only Sitka spruce could have become established there. Accordingly, we might hypothesize that the Alaskan Gulf coast climate during last 4000 years has become drier and that this has allowed the spread of the two hemlock species.

However, this solution is most likely too simple, as it fails to take into account the myriad of other ecological conditions needed to foster the growth and development of hemlock. For instance, Sitka spruce is not very tolerant of shade, whereas the two hemlock species are very shade tolerant. Perhaps hemlock successfully competes with other tree species only in regions where other trees (in this case, spruces) are already well established and growing tall. In any case, we do not yet know enough about the factors that control the spread of these trees across the Alaskan landscape.

Perhaps the lesson to be learned from all of this is that the Pacific coastal forest in Alaska is a very dynamic ecosystem that has only recently reached its modern form. It did not race along the Gulf coast of Alaska after the last ice age. Rather, certain elements of the ecosystem were kept at bay for many thousands of years. Alaskan foresters are also quick to point out that the current ecosystems in this region are far from stable. Species of trees, as well as other organisms in the ecosystem, are still shifting distributions, invading new regions and abandoning old ones. The disturbances brought on by the logging industry constitute just the most recent force to be added to the ecological equation. No one knows in what direction the coastal forest ecosystem will be driven by this latest, human disturbance. All we can say with certainty is that by the time we come to understand human impacts on these majestic forests, it will probably be beyond our powers to restore the ecosystem to its pristine state, at least during any time scale that humans can appreciate.

Suggested Reading

Ager, T. A. 1983. Holocene vegetation history of Alaska. In Wright, H. E., Jr. (ed.), *Late Quaternary Environments of the United States, Volume 2, The Holocene.* Minneapolis: University of Minnesota Press, pp. 128–141.

Hamilton, T. D., and Thorson, R. M. 1983. The Cordilleran ice sheet in Alaska. In Wright, H. E., Jr. (ed.), *Late Quaternary Environments of the United States, Volume 1, The Late Pleistocene.* Minneapolis: University of Minnesota Press, pp. 38–52.

Heusser, C. J. 1985. Quaternary pollen records from the Pacific Northwest coast: Aleutians to the Oregon–California boundary. In Bryant, V. M., Jr., and Holloway, R. G. (eds.), *Pollen Records of Late-Quaternary North American Sediments.* Austin, Texas: American Association of Stratigraphic Palynologists, pp. 141–164.

Karlstrom, T. N. V. 1964. *Quaternary Geology of the Kenai Lowland and Glacial History of the Cook Inlet Region, Alaska.* United States Geological Survey Professional Paper 443. Denver, Colorado: U.S. Geological Survey. 69 pp.

Miller, D. W. 1984. *A Guide to Alaska's Kenai Fjords.* Seward, Alaska: Wilderness Images. 94 pp.

Wiles, G. C., and Calkin, P. E. 1993. Neoglacial fluctuations and sedimentation of an iceberg-calving glacier resolved with tree rings (Kenai Fjords National Park, Alaska). *Quaternary International* 18:35–42.

8

GLACIER BAY NATIONAL PARK
A Landscape in Motion

The story of the European discovery of Glacier Bay sums up the grandeur, sense of wonder, and potential danger of the place as well as any modern account could do. Dave Bohn tells this story in its entirety, but a brief summary is sufficient for our purposes. The first European visitors, Vitus Bering and Alexis Tchirikov, mounted a naval expedition there under the Russian flag in 1741. Tchirikov arrived a day before Bering and led a party of explorers, who set out in a small boat from their ship, the St. Paul, on July 18. They made for Lisianski Strait to enter Glacier Bay. The landing party was never heard from again.

In 1879, naturalist John Muir came to Glacier Bay in a dugout canoe in search of glaciers. He was accompanied by Presbyterian missionary S. Hall Young, Toyatte, a Stikeen tribal chief, Kadichan, son of a Chilkat chief, and two other natives. They paddled about 250 miles, from Fort Wrangell to Glacier Bay. They began their voyage on October 14 and arrived at Berg Bay on the night of October 24. Although Muir was not the first European to see Glacier Bay, he was the first to publish his discoveries and make it famous.

Not surprisingly for that time of the year, the weather at Glacier Bay was horrendous. The small party was met by icy rain and wind. They also met a group of residents, members of the Hoonah tribe, who offered them shelter from the storm and informed them that this place was called Sit-a-da-kay, or Ice Bay. They informed Muir that there were indeed great glaciers and mountains in the vicinity, even though they were hidden from view by the clouds. They said that the greatest of the ice mountains was at the head of the bay, their favorite hunting ground for seal. The next morning, Muir climbed up the ridge just east of what is now Charpentier Inlet and beheld Glacier Bay for the first time. He recorded his thoughts:

> I reached a height of fifteen hundred feet, on the ridge that bounds the second of the great glaciers. All the landscape was smothered in clouds, and I began to fear that as far as wide views were concerned I had climbed in vain. But at length the clouds lifted a little, and beneath their gray fringes I saw the berg-filled expanse of the bay, and the feet of the mountains that stand about it, and the imposing fronts of five huge glaciers, the nearest being immediately beneath me. This was my first general view of Glacier Bay, a solitude of ice and snow and newborn rocks, dim, dreary, mysterious. I held the ground I had so dearly won for an hour or two, sheltering myself from the blast as best I could, while with benumbed fingers I sketched what I could see of the landscape, and wrote a few lines in my notebook. Then, breasting the snow again, crossing the shifting avalanche slopes and torrents, I reached camp about dark, wet and weary and glad. (John Muir, 1915, pp. 144–145)

Glacier Bay is a powerful landscape. The mighty glaciers, backed by towering, ice-covered peaks, dwarf the human visitor. In this place, perhaps as in no other in North America, we can catch a glimpse of the ancient Pleistocene world, when mighty glaciers smothered now-familiar landscapes from Seattle to Manhattan.

Modern Setting

When the abundant moisture made available by the Gulf of Alaska meets tall mountains of the St. Elias range at Glacier Bay, the result is equally abundant precipitation. At the coast, this falls mostly as rain; on the mountains, it falls as snow, up to 4.6 m (15 ft) per year. This snow is the source of the glaciers that flow down to the sea. The vertical relief of the mountains is staggering. The highest point in the park, Mount Fairweather, rises 4663 m (15,300 ft) above sea level, yet the peak is only 24 km (15 miles) inland from the bay.

The glaciers feeding into Glacier Bay are constantly shifting: advancing, retreating, and changing course. The forces that govern the size of the glaciers in this region are many and complex but can be summed up in three main factors:

temperature, precipitation, and the slope and aspect of the mountains on which they are born. During the Pleistocene, temperatures were low enough to foster the growth of huge glaciers, which came together to form ice sheets covering most of southern Alaska. Today, temperatures favorable for the growth and maintenance of glaciers are found only on the high mountains of this region. At one level, the equation is relatively simple: If the climatic balance between annual snowfall and snow melt is tipped toward cold conditions, the ice increases in thickness and expands at the **glacial terminus.** If the scale tips toward warmer conditions (that is, more snow melts in summer than accumulates in winter), a glacier starts to thin, and its terminus melts back. As William Boehm points out, however, there are other factors that influence the size of glaciers, including surface winds, solar radiation, cloud cover, reflectivity of the ice, and physical properties of the ice itself. For instance, in regions where glacial ice is covered with a thick layer of pure, white snow, most of the sun's energy is reflected off the surface, so little melting takes place. On the other hand, glacial regions with a thin mantle of dark rocks, soil, and other debris soak up significant amounts of solar energy and melt more rapidly.

The modern geography of Glacier Bay is far from static. Glacial retreat and advance continually change the landscape. One example is the recent history of Grand Pacific Glacier at the back of Tarr Inlet (Fig. 8.1). From about 1895 to 1970, the glacier receded about 1.6 km (1 mile) per year until its terminus was very near the international border between Alaska and British Columbia. Canadian geographers were all set to declare a new saltwater port for their country, on Tarr Inlet. They went so far as to plant a Canadian flag near the glacial terminus during the summer of 1974. However, the glacier did not cooperate. To the consternation of the Canadians, it had stopped receding, and at the time they planted their flag, the terminus was at least 1.3 km into Alaskan territory.

Geologic and biological processes are sped up in Glacier Bay, with new moraines appearing every few years, and newly exposed mineral soils supporting first mosses and lichens and then, within a few decades, a vegetational succession that leads to Pacific coastal forest. Two hundred years ago (a blink of an eye in geologic history), Glacier Bay was buried in ice more than 1200 m (4000 ft) thick. This ice cover built up during the "Little Ice Age," a time of unusually cold, wet weather throughout the northern hemisphere, which lasted from the sixteenth through the eighteenth century.

The National Park lies within a large, semicircular rim of mountains, the Fairweather Range. The Brady Icefield extends some 50 km along the eastern slope of the mountains. The mountains and icefields play a large role in shaping the regional weather patterns. Although the climate along the coast is nearly as mild as that of Seattle or Vancouver, the more northerly latitude of Glacier Bay dictates

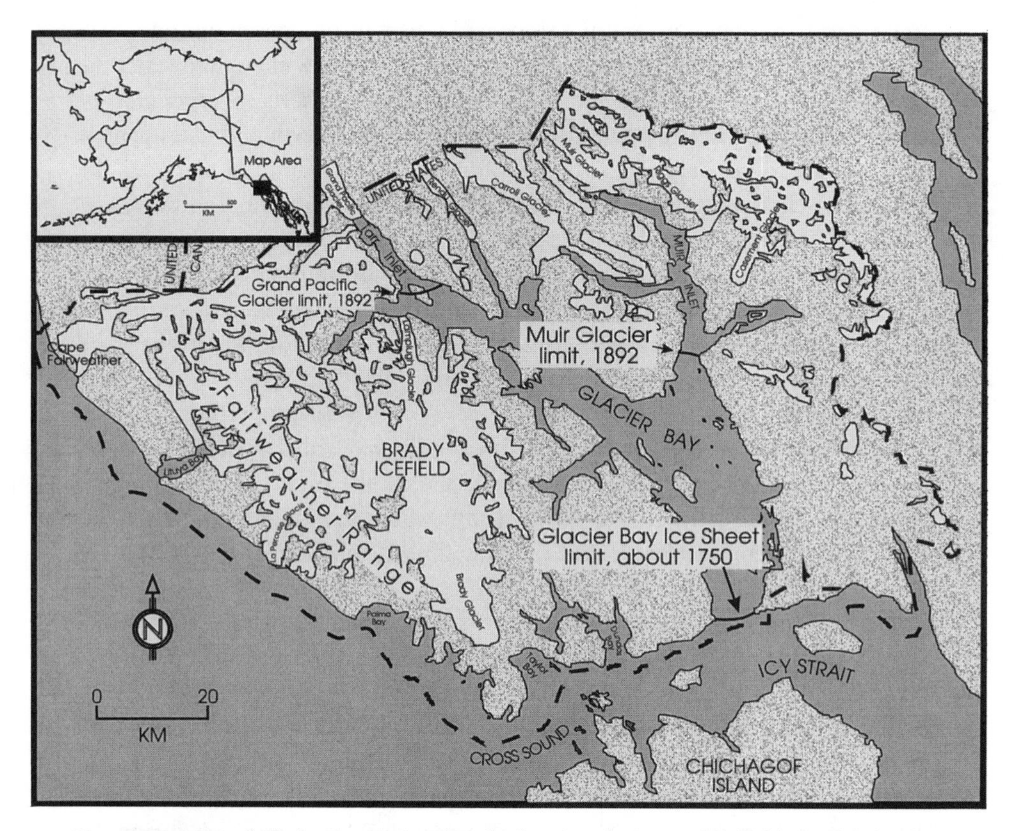

Figure 8.1. Map of Glacier Bay National Park, showing glaciers and icefields (park boundary shown by dashed line).

mountain climates that are much more severe. The steep climatic gradient of the mountain slopes is also reflected in the shortened vegetation zones of this region, summarized in Figure 8.2. Regions below about 800 m elevation that are free of ice are forested by tall trees in very dense stands. Starting at the bottom, the forests growing near tidewater are dominated by Sitka spruce. Unglaciated terrain up to 500 m elevation is clothed in Pacific coastal forest and muskeg. The coastal forest in this region is dominated by western hemlock, Sitka spruce, and Alaska yellow cedar. Treeline is only about 790 m (2600 ft) above sea level in the park. The trees growing near treeline are mostly spruces. These trees take on the characteristic **krummholz** form of low, twisted trunks and bows hugging the ground. Above treeline, a mixture of arctic and alpine tundra plants are found on gentler slopes and alpine meadows (800 to 900 m, or 2620 to 2950 ft above sea level), while steeper slopes and regions above about 1600 m (5250 ft) elevation support only sparse, patchy fell field vegetation. Most high-altitude slopes are either bare rock or covered with ice.

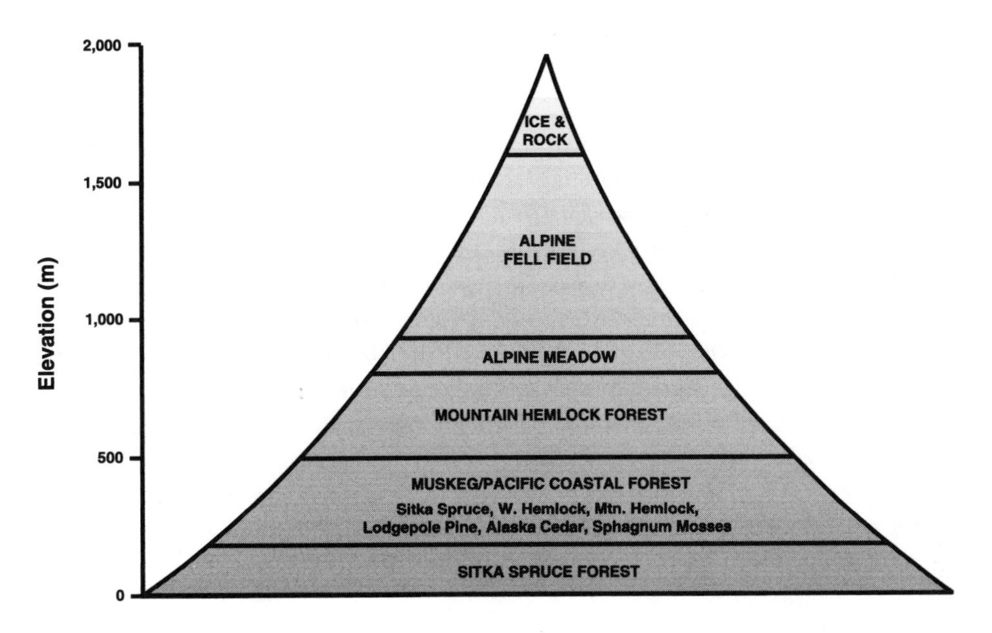

Figure 8.2. Modern vegetation zones of southeastern Alaska. (After Barnowsky et al., 1987.)

The forests of Glacier Bay are tenacious, taking hold in seemingly impossible places. Some stands of trees are even found growing on the glaciers themselves. For instance, Cape Fairweather, at the northwest corner of the park, is the terminus for Fairweather Glacier, which has stands of spruce growing in the layer of rocks and soil covering the outermost parts of the glacier. The rocks and soil were churned up near the glacier's edge as it advanced.

The animal life in Glacier Bay National Park is essentially that of the entire Alaskan Gulf coast. Mountain goats keep to the alpine zone, where they feed on tundra vegetation. Wolves, coyotes, and wolverines are found in many habitats (generalist predators like these are free to wander far and wide in search of prey). The spruce-hemlock forests are home to a wide variety of wildlife, including red squirrels, flying squirrels, red-backed voles, and black and grizzly bears. Sea otters were once quite plentiful along the Alaskan Gulf coast, but they were extensively hunted in the Glacier Bay region, beginning with the Russians in the eighteenth century. Overhunting caused them to be exterminated in Glacier Bay. These otters have been reintroduced to the park and have become successfully established in the Icy Strait region, according to Fish and Wildlife Service studies. They are expected to reenter Glacier Bay in the near future. River otters have persisted in this region, and healthy populations remain there.

The marine mammals that frequent Glacier Bay include harbor seals, Stellar sea lions, porpoises, and whales, including humpbacks and killer whales. Harbor seals give birth to their young on floating icebergs calved from tidewater glaciers in the bay. Humpback whales are seasonal visitors to southeastern Alaska. They come to harvest the rich crops of invertebrates that grow here in summer months. The killer whales prey mostly on fish; some spend just part of the year in southeastern Alaska, and others migrate south during the winter, to the coasts of British Columbia and beyond. The migration patterns of the bird and mammal species that currently spend their summers in Alaska may have formed long ago, perhaps in the Pleistocene. In North America, northern summer migrations would likely have been made to Beringia, because it was the only unglaciated high-latitude region available during glacial intervals of the Pleistocene. Scientists are only beginning to try to find ways to reconstruct this part of the history of animal life in North America.

Geologic History

Pleistocene glaciation was extensive throughout southeastern Alaska, as evidenced by the many glacial landforms left behind. Glacial ice scoured valleys into the characteristic U-shape, carved fjords, and trimmed peaks into pyramid shapes. As the glaciers retreated, they dropped vast amounts of debris, ranging from building-sized boulders to silt and clay. Many moraines can be seen on the landscape; many more are now underwater, on the continental shelf regions of the Gulf of Alaska. According to geologist Bruce Molina, glacial erosion of the continental shelf in the gulf produced seven large sea valleys, cut into the shelf during periods of lowered sea level. Some of those valleys were filled with ice as recently as 12,000 years ago.

Curiously, it appears that parts of the Alaskan Gulf coast were not glaciated during the late Wisconsin interval. For instance, geologic evidence compiled by Dan Mann suggests that Lituya Bay and adjacent coastal regions on the west side of Glacier Bay National Park were ice-free during the late Wisconsin. The limited extent of glacial ice in this region was probably the result of two factors. First, the catchment areas of local mountain glaciers are relatively small, and the mountain slopes facing the Pacific coast are quite steep. This configuration does not allow the type of massive ice buildup required for expansion of a small mountain glacier into a large regional glacier or ice sheet. Second, the glaciers in this region front on deep water along an open coast, where iceberg calving was probably always very active during periods of relatively high sea level, or when the continental shelf was covered by floating ice shelves, which allowed sea water to reach the fronts of tidewater glaciers.

Earlier in the Wisconsin, glaciers may have been more extensive. At Lituya Bay, peat beds of last interglacial age are overlain by **glacial outwash** and **till** deposits, which were probably deposited during an early Wisconsin Glaciation. The late Wisconsin ice cover over southeastern Alaska was undoubtedly quite extensive, even if some regions remained ice-free because of local topographic constraints. Deglaciation was well under way in this region by 14,000 yr B.P. Apparently, most lowland regions were free of ice shortly thereafter, but peats and lake sediments rich in organic matter did not begin accumulating until about 11,000 yr B.P. The postglacial vegetation record begins at that time.

During the Holocene, tongues of ice advanced onto the continental shelf of the Gulf of Alaska several times. The most recent of these advances took place in the early 1900s. Glaciers fronting Lituya Bay had their maximum Holocene advance during the last 600 years (Fig. 8.3). Glacially scoured troughs and depressions can be seen adjacent to many of today's coastal glaciers. Such depressions are associated with LaPerouse, Fairweather, and Grand Plateau glaciers, suggesting that these glaciers advanced onto the shelf during the Holocene. However, geologists have yet to unravel the extent and timing of the advance of the Glacier Bay Ice Sheet onto the shelf.

The sediment load left by retreating tidewater glaciers is extremely high. Many of the coastal embayments along the Gulf of Alaska accumulate a thickness of more than a meter of new glacially derived sediment per year. Studies in Muir Inlet, Glacier Bay, documented the deposition of more than 10 m of glacial sediment during the first year following the retreat of the tidewater Muir Glacier (Fig. 8.4).

The landscape comprising Glacier Bay National Park is constantly changing because of glacial activity (Fig. 8.5). This statement applies as much to the modern environment as to the Pleistocene. Whereas most regions of Alaska could be said to have "settled down" a bit since the last ice age, Glacier Bay is as dynamic as ever.

Development of Postglacial Ecosystems

The postglacial vegetation history of the Glacier Bay region (Fig. 8.6) has been summarized by Ager, by Heusser, and by Barnowsky and colleagues. The earliest records of postglacial vegetation begin with ponderosa pine parkland, recorded in the basal pollen assemblages from Adams Inlet, an eastern arm of Glacier Bay

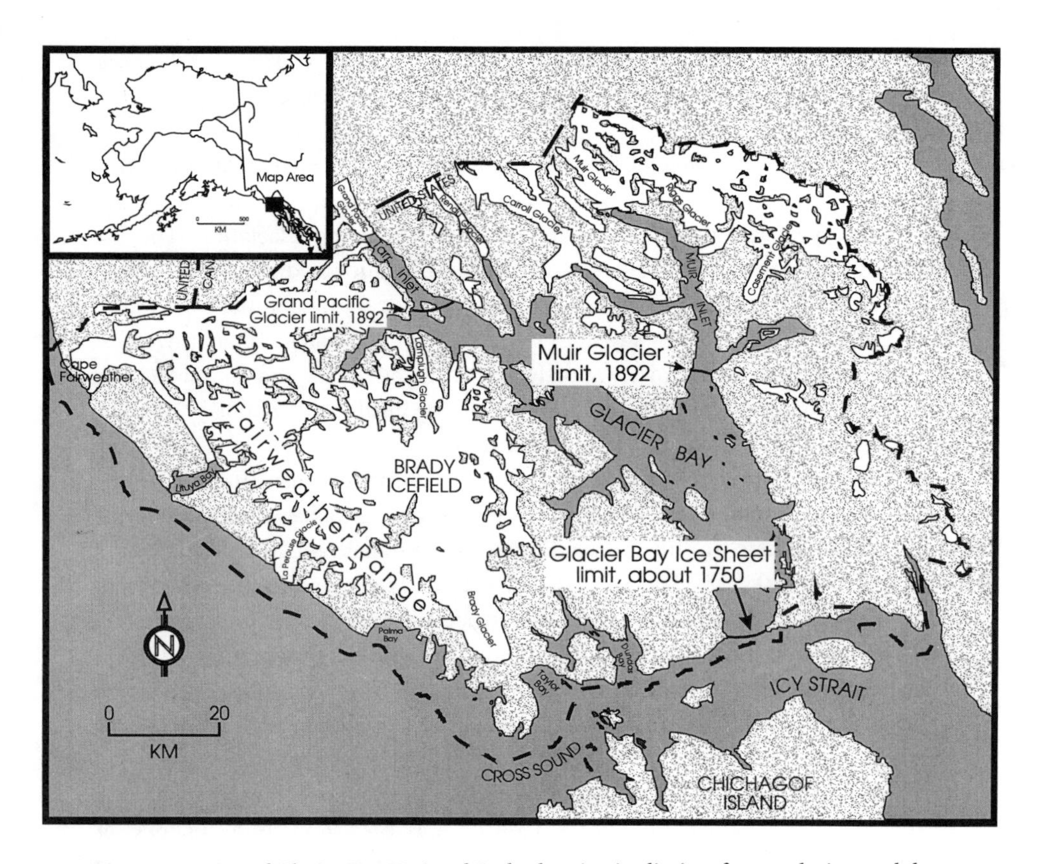

Figure 8.3. Map of Glacier Bay National Park, showing ice limits of some glaciers and the Glacier Bay Ice Sheet within the last 250 years. (After Boehm, 1975.)

(Fig. 8.7). This basal assemblage, from the Forest Creek Formation, has been dated by Ager at 10,900 yr B.P. The oldest pollen assemblage from Montana Creek, near Juneau (Fig. 8.7), has been dated at 10,800 yr B.P. The early vegetation at Montana Creek was a lodgepole pine-alder parkland. This same vegetation was indicated by pollen from basal assemblages at Muskeg Cirque, near Lituya Bay (Figs. 8.6 and 8.7). It is reasonable to assume that the initial colonization of bare mineral soil at these sites began with mosses, lichens, and herbs and that the establishment of trees took place many centuries later. However, we know nothing of this successional history beyond what has been observed in modern times.

At Adams Inlet and Montana Creek, forests of Sitka spruce, western hemlock, and alder formed by about 8500 yr B.P. Sitka spruce invaded the Lituya Bay region by 10,300 yr B.P. However, western hemlock did not appear in the pollen records from Muskeg Cirque until about 7000 yr B.P. The early Holocene forests of the

Figure 8.4. The terminus of Muir Glacier in 1983. This glacier has virtually finished a drastic retreat across Muir Inlet, caused by a tidewater glacier instability involving the interaction between the rate of iceberg calving and water depth at the terminus. The terminus is now in shallow water, the rate of iceberg calving is low, and little or no further retreat is expected. (Photograph by Dr. Mark F. Meier, reprinted by permission of Dr. Meier.)

Glacier Bay region resembled the modern forests of the region, but the complete modern forest composition did not come together until later. In the meantime, glaciers fronting Glacier Bay began advancing, and by about 7500 yr B.P., many of the region's forests were buried by ice. This ice cover persisted in some regions for more than 3000 years, right through the middle of the Holocene. The broken stumps of early Holocene trees can still be seen, poking through late Holocene moraines along Adams and Muir Inlets (Fig. 8.8).

Establishment of Modern Ecosystems

By about 4000 yr B.P., most regional glaciers had retreated again, and mountain hemlock invaded the Glacier Bay district. Modern forest composition, including the two species of hemlock, Sitka spruce, and lodgepole pine, was established at Montana Creek and Adams Inlet by about 1500 yr B.P. As in the Kenai Fjord scenario, we are left with an interesting puzzle concerning this late establishment of modern forest types. Why is it that some species failed to colonize these regions until so recently, whereas other species with seemingly similar ecological requirements got in thousands of years earlier? Palynologist Dorothy Peteet studied the

Figure 8.5. (A) The Muir Glacier in 1892, photographed by H. F. Reid. The 1972 terminus position is marked by an arrow. (B) The Muir Glacier in 1965, photographed by Austin Post. Arrow points to the same location, the 1972 terminus position. Between 1892 and 1965, the Muir Glacier retreated 32 km. (Photographs reproduced from Brown et al., 1982.)

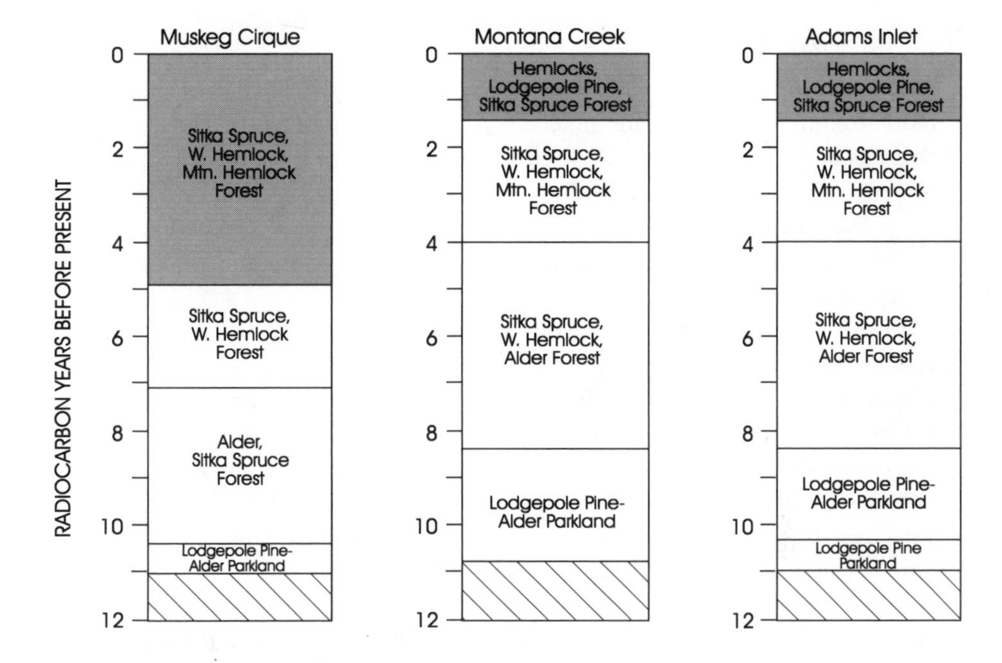

Figure 8.6. Summary diagram of the postglacial development of plant communities at three localities in or near Glacier Bay National Park. Muskeg Cirque data after Barnowsky et al. (1987); Montana Creek and Adams Inlet data after Ager (1983). Shaded zones indicate the intervals of modern vegetation.

vegetation history of the Munday Creek site, near Icy Cape, where the two hemlock species typical of the Pacific coastal forest arrived between 3800 and 3500 yr B.P. Her theory is that the spread of hemlocks was linked with increases in storm frequencies and precipitation.

The comings and goings of lodgepole pine in southeastern Alaska pose another interesting ecological problem. In Alaska, this species reaches its northern limit near Haines, at the northern end of the Lynn Canal. Why was it so successful in the early postglacial vegetation around Glacier Bay, only to drop out of sight until the last few centuries? These problems would be difficult enough to solve if each species acted independently of all other species (that is, if each species were responding only to changes in the physical environment), but, unfortunately for the poor ecologists, all of these species interact with many other species in a biological community. Perhaps careful reconstructions of past communities will help us come to understand how the current ecosystems work.

Ecological Succession in a Glaciated Landscape

Glacier Bay offers ecologists tremendous opportunities to study the succession of biotic communities from bare mineral soil through mature forest. The reason is that this biological drama is reenacted in the park on a year-by-year basis, as the various glaciers advance over old forests and retreat again, exposing bare mineral soils. The scouring action of glacial ice acts as a giant eraser on the biological chalkboard, sweeping it clean and allowing some new writing to be done, then sweeping it clean again.

Ecologists have been studying this process at Glacier Bay for most of this century. Starting with Cooper in 1923, a series of studies have been made. One good summary is found in an article by William Reiners and others. The bare soil left by the receding glacier is colonized first by mosses and lichens, then by sedges and horsetails. Within 5 to 20 years, mountain avens becomes dominant and forms a thick mat, which nurtures the seedlings of willows, birches, balsam poplar, and alders. By 20–40 years into the succession, these have formed a dense shrub thicket dominated by alder. Anyone who has tried to hike through one can testify to the density of these alder thickets! Soon cottonwoods and Sitka spruce are growing up taller than the surrounding alder thicket, and by 50 to 70 years into the sequence, these two tree species have become important in the plant community (50% cover by spruce and cottonwood). By 75–100 years Sitka spruces become dominant and form a spruce forest. By this time, the forest floor has become a thick carpet of

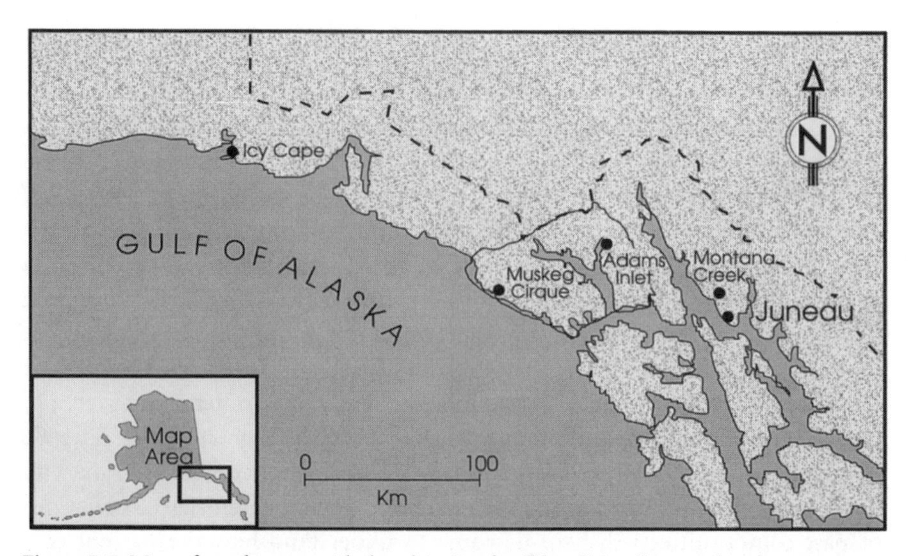

Figure 8.7. Map of southeastern Alaska, showing fossil localities discussed in the text.

Figure 8.8. Stumps of early Holococene trees, exposed by the retreat of the Morse Glacier, 1906. (Photograph by C. W. Wright, from U.S. Geological Survey Photo Library.)

mosses. This substrate, and the shade provided by the spruces, allows the invasion of mountain hemlock and western hemlock. As the forest matures during the next century, hemlock increases at the expense of Sitka spruce. The final step in this succession, at least on level ground or gentle slopes, is the gradual buildup of sphagnum mosses, creating the muskeg community. The mosses form such a dense mat in some localities that they choke the roots of the trees, robbing them of nutrients. The trees may eventually die, leaving a landscape of dead snags and wet sphagnum. As unsavory as it may seem, the most stable plant community of the system, the ultimate end of succession, is often a half-dead forest choked with sphagnum. It is as if Nature had made a decree, "From moss thou hast come and to moss thou shalt return."

Continued studies of plant succession of deglaciated landscapes in the Glacier Bay region have shown that the pattern of succession varies considerably from locality to locality and is not as predictable as it seemed to earlier workers. Seed availability appears to be the overriding factor in the process. For instance, spruces enter the picture much more slowly at the upper end of the bay than at the lower end because of the difficulty in seed transport up the mountain slopes.

Between the dynamics of the glaciers and the rapid responses of the forests, Glacier Bay is a landscape in motion. One of the troubles mentioned in John Muir's accounts of his travels there with the Indians in the last century was this: his Indian guide had not been there for many years, and the place had changed so much he did not recognize any landmarks. The National Park created to protect this ever-changing piece of the Alaskan wilderness may fascinate us, challenge us, or overwhelm us, but it will never be boring.

Suggested Reading

Ager, T. A. 1983. Holocene vegetation history of Alaska. In Wright, H. E., Jr. (ed.), *Late Quaternary Environments of the United States, Volume 2, The Holocene.* Minneapolis: University of Minnesota Press, pp. 128–141.

Barnowsky, C. W., Anderson, P. M., and Bartlein, P. J. 1987. The northwestern U.S. during deglaciation: Vegetational history and paleoclimatic implications. In Ruddiman, W. F., and Wright, H. E., Jr. (eds.), *Geology of North America, Volume K-3, North America and Adjacent Oceans During the Last Deglaciation.* Boulder, Colorado: Geological Society of America, pp. 289–321.

Boehm, W. D. 1975. *Glacier Bay.* Anchorage, Alaska: Alaska Northwest Publishing Company. 134 pp.

Bohn, D. 1967. *Glacier Bay: The Land and the Silence.* San Francisco: The Sierra Club. 165 pp.

Brown, C. S., Meier, M. F., and Post, A. 1982. *Calving Speed of Alaska Tidewater Glaciers, with Application to the Columbia Glacier.* U. S. Geological Survey Professional Paper 1258-C. Denver, Colorado: U.S. Geological Survey. 13 pp.

Cooper, W. S. 1923. The recent ecological history of Glacier Bay, Alaska. *Ecology* 4:93–128, 223–246, 355–365.

Heusser, C. J. 1985. Quaternary pollen records from the Pacific Northwest coast: Aleutians to the Oregon–California boundary. In Bryant, V. M., Jr., and Holloway, R. G. (eds.), *Pollen Records of Late Quaternary North American Sediments.* Austin, Texas: American Association of Stratigraphic Palynologists, pp. 141–164.

Mann, D. 1983. Glacial history near Lituya Bay, Alaska. In Thorson, R., and Hamilton, T. D. (eds.), *Glaciation in Alaska.* Alaskan Quaternary Center, University of Alaska Museum, Occasional Paper No. 2. Fairbanks, Alaska: University of Alaska, pp. 62–66.

Molina, B. F. 1983. Late Wisconsinan and Holocene glaciation of the Alaskan continental margin. In Thorson, R., and Hamilton, T. D. (eds.), *Glaciation in Alaska.* Alaskan

Quaternary Center, University of Alaska Museum, Occasional Paper, No. 2. Fairbanks, Alaska: University of Alaska, pp. 67–70.

Muir, J. 1915. *Travels in Alaska.* Cambridge: Houghton, Mifflin & Co.

Reiners, W. A., Worley, I. A., and Lawrence, D. B. 1971. Plant diversity in chronosequence at Glacier Bay, Alaska. *Ecology* 52:55–69.

Viereck, L. A., and Little, E. L. 1986. *Alaska Trees and Shrubs.* U.S. Department of Agriculture Forest Service Agriculture Handbook No. 410. Washington, D.C.: U.S. Department of Agriculture. 265 pp.

9

CONCLUSION

Alaska is a fascinating place, made all the more interesting because of its unique history during the Pleistocene. While most high-latitude regions of North America were covered with ice during the major glaciations, large parts of Alaska were ice-free, including the interior regions north of the Alaska Range and the western regions of the state.

Alaska and unglaciated Siberia were connected by a land bridge during lengthy intervals of the Pleistocene, and together these regions formed the land we call Beringia. Beringia was just about the only refuge for cold-adapted plants and animals that was north of the huge continental ice sheets of Eurasia and North America.

At the end of the last glaciation (about 12,000 years ago), the zone of tundra that existed south of the Laurentide and Cordilleran ice sheets was obliterated, and the cold-adapted biota living in this narrow pocket of tundra was essentially wiped out. This occurred because the southern edges of the ice sheets retreated too slowly to allow the biota to migrate north. The elimination of cold-adapted species from regions south of the ice sheet meant that the biota that survived the Wisconsin Glaciation in the Beringian refuge became practically the only source of plants and animals for the recolonization of arctic and subarctic regions of Canada when those regions were finally deglaciated.

During the last glaciation, Beringia supported a rich, diverse fauna of large mammals, including grazers, browsers, predators, and scavengers. Many of these megafaunal mammals became extinct at the end of the Pleistocene (about 11,000 years ago), but a few live on, including the musk ox, caribou, Dall sheep, saiga antelope (now only in Asia), grizzly bear, and arctic fox. Most paleoecologists believe that a wide variety of plant communities developed in Beringia to support such a variety of large mammals. The vegetation mosaic was undoubtedly complex over this huge region with all its topographic settings. The most important of these communities was probably the steppe-tundra, a curious mixture of arid grassland and arctic tundra species that no longer exists in the same form it had during the Pleistocene. What seems clear from the fossil mammal evidence is that Beringia was not vegetated with sparse polar desert or fell field plants except in isolated regions that experienced extremes of climate or rocky soils.

The role of humans in the extinction of the North American megafauna remains unclear, as does the timing of humans entering the New World. However, if the first people entered North America near the end of the Wisconsin Glaciation, it seems unlikely that the simultaneous extinction of so many large mammals was mere coincidence.

How did the first human populations enter the New World? The most widely held theory among archaeologists is that they entered via the Bering Land Bridge some time near the end of the Wisconsin Glaciation. However, the strong evidence for human occupation of southern South America as early as 13,000 yr B.P. may force a new model to the front. The fact that Monte Verde, Chile was occupied that early may mean that Asian peoples came south by boat, along the Pacific coast of the Americas.

New discoveries are always cropping up in paleoecology, challenging old theories and forcing us to rethink our assumptions. Scientific discoveries often take years of painstaking work to accomplish; at times, however, these discoveries are revolutionary, knocking large holes in old theories with one blow. For instance, as I was writing this book, the Russian paleontologists Vartanyan et al. (1993) published startling results of their work on Wrangel Island, off the northeastern coast of Siberia (Fig. 9.1). These researchers obtained numerous radiocarbon dates on mammoth bone collagen, showing that the last mammoths died out on Wrangel Island only about 3700 yr B.P. No other mammoth remains have ever been found that were younger than about 9500 yr B.P.

Wrangel Island is unique in several ways. It was connected to mainland Siberia until about 12,000 yr B.P., when the intervening continental shelf was flooded by rising sea level. The modern climate and vegetation of the island closely resemble steppe-tundra environments of Pleistocene Beringia. Also, humans did not enter Wrangel Island until the late Holocene. If humans had been there earlier, they

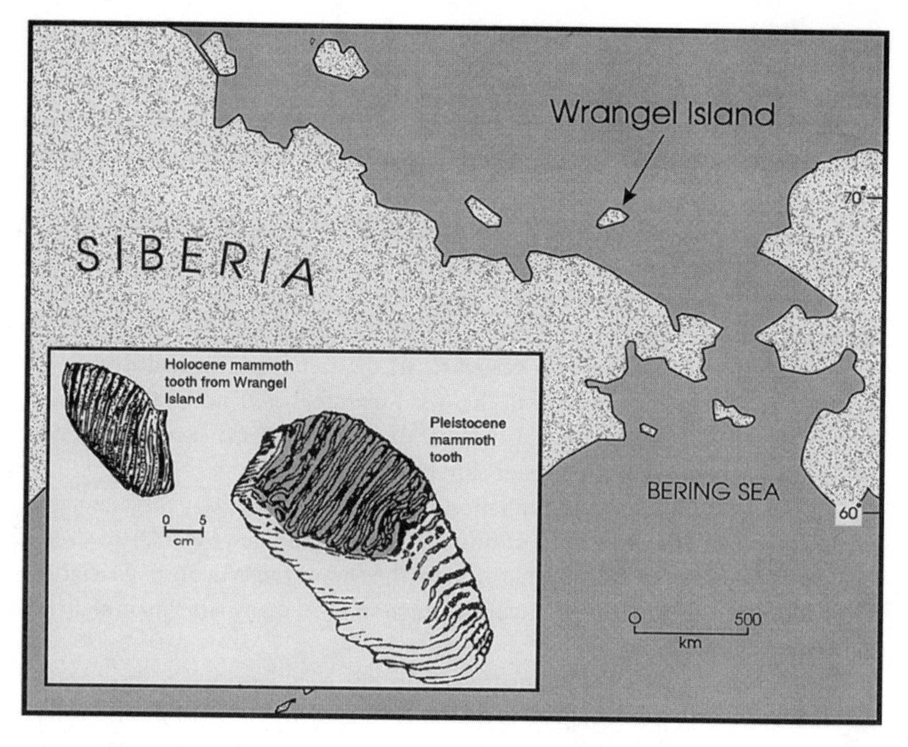

Figure 9.1. Map showing location of Wrangel Island off the northeastern coast of Siberia. Inset shows last molar tooth of dwarf mammoth, radiocarbon dated at about 4000 years B.P. (after Vartanyan et al., 1993), compared with Pleistocene mammoth molar.

might have hunted the mammoths to extinction. The mammoths that survived there were substantially smaller in size than their Pleistocene ancestors (see the mammoth tooth illustrated in Fig. 9.1). This is typical of mammal populations isolated on small islands. The Wrangel discovery is just one example of how old, seemingly well-established ideas about Beringia have to be discarded with painful regularity.

While Eastern Beringia (Alaska and unglaciated Yukon Territory) enjoyed ice-free conditions, most of southern Alaska was locked in ice during the Wisconsin Glaciation. Glaciers descending from the Alaska Range, the Chugach Mountains, the Wrangel–St. Elias Mountains, and other ranges coalesced along various fronts, forming thick sheets of ice. However, there is tantalizing evidence that some regions in the south remained ice-free. The tops of some high mountains may have been nunataks, refuges for arctic and alpine biota. Ice-free pockets along the southeastern shores of the Alaskan Gulf coast may have harbored lowland biota. Much remains to be learned before these speculations become hard evidence for regional glacial refugia.

The climate of southern Alaska remained cool and wet enough to maintain mountain glaciers even through the warmest parts of the Holocene. Advancing glaciers in the mid-Holocene buried coastal forests that had been established for thousands of years. During the last thousand years, glaciers in southeastern Alaska came together to form an ice sheet that covered Glacier Bay. Many sites around Glacier Bay have come free from the ice only during the twentieth century.

The combination of tectonic action and Holocene glaciation has produced some very dynamic landscapes in southern Alaska. Movements of glaciers have blocked streams, creating lakes that lasted a few years until the water worked its way through the base of the ice. Some regions have been subsiding into the sea, a gradual process occasionally quickened dramatically by earthquakes. Shallow valleys became fjords. Inlets have opened up and then once more became choked with ice. In southern Alaska, very little remains the same from decade to decade, much less for thousands of years.

Southern Alaska is an open-air laboratory for glacial geologists and succession ecologists. The geologists have discovered, among other things, that the movements of tidewater glaciers are not always in step with regional climate change or with glaciers that are land based. The ecologists have come to the conclusion that the ultimate plant community for many lowland regions is a muskeg choked with sphagnum in which many if not most of the trees are dead or dying.

Alaskan Wilderness: The Challenges and Rewards

Alaska is treasured by residents, tourists, and scientists alike because it remains so unspoiled. Raw, powerful nature is preserved here as in few other places left on Earth. Scientific field work in Alaska is enormously expensive, often physically challenging, and sometimes dangerous to life and limb. Helicopter charter service in Alaska currently costs about $850/hour. There is nothing quite as lonely as waving goodbye to that high-priced pilot as the helicopter climbs back into the sky, leaving the passengers stranded on an all-too-isolated river bank. But you are never alone in the Alaskan "bush." There are always plenty of mosquitoes, black flies, midges, punkies, and no-see-ums to keep you on your toes.

Field work in Alaska is usually limited to a few brief weeks of summer. Of course, as many researchers have observed, it never really gets dark in Alaska during June, July, and August, so you can work around the clock to get things done. However, even unlimited daylight does not make the rain stop in southeastern Alaska or cause the damp mist and fog to depart from the arctic coast near Barrow. Then again, there is nothing like the challenge of clinging to a steep, 50-m-tall riverbank when it starts to rain and the silt underfoot takes on the consistency of cooking oil.

Nevertheless, most of us who do field work in Alaska are drawn back again, year after year, as often as we can scrape up enough money to mount an expedition. We keep coming because Alaska is the most horrendous, exasperating, exhausting place we've ever fallen in love with. The wildness, the untamed feel of the place, gets a hold on you. All the rain-drenched afternoons and the mosquito bites are a small price to pay for the privilege of seeing a pair of grizzly cubs frolic on a tundra knoll or listening to a pack of wolves howl across a valley or tasting sun-ripened blueberries from a boggy trail or digging a trowel into sediments jam-packed with 500,000-year-old fossils kept fresh for you by the permafrost's deep freeze.

I've been fortunate enough to spend a great deal of time in Alaska; for the better part of a decade I've been going there to collect insect fossils. I hope to be there for many more field seasons and maybe even a few more winters. (There is nothing like trying to get a car started at 40° below zero to give you a deeper appreciation for the hardiness of the first Paleoindians who passed through the country.)

Developing the Historical Perspective

My hope in writing this book has been that the reader will catch some of the sense of wonder and fascination for the Alaskan wilderness, as preserved in her national parks. But more than that, I hope that you will come to understand that what can be seen there today, however awe-inspiring, is only the last page in the very long book of life as written for this great region. That book is a strange one for most, written in pages of peat, silt, and organic muck. Because it is unfamiliar, most people have overlooked it. But the ancient past has a lot to offer, once you have been introduced to it.

When you have a historical perspective (some might call it a prehistoric perspective), you begin to see nature in a new way. This perspective is essentially an appreciation that modern ecosystems and landscapes have been shaped by past events. The current crops of plants and animals in a biological community are just the biological actors on stage for one act of a very long play. Furthermore, the combination of species in an ecosystem may or may not be the ones best suited for that particular environment. Some will probably be gone in a few centuries. Newcomers, better fit for the system, will squeeze them out. The scene appears stable to our eyes, but in reality it is constantly shifting. So come to Alaska's great wilderness, stand on a high promontory, as John Muir did at Glacier Bay, look over the landscape, and ask yourself, "How did it come to be this way?"

Suggested Reading

Vartanyan, S. L., Garutt, V. E., and Sher, A. V. 1993. Holocene dwarf mammoths from Wrangel Island in the Siberian arctic. *Nature* 362:337–340.

GLOSSARY

Alkaline sediments Sediments that contain more hydroxyl ions than hydrogen ions; sediments with a pH greater than 7.0.

Acidic sediments Sediments that contain more hydrogen ions than hydroxyl ions; sediments with a pH less than 7.0.

Alpha (α) particle A helium nucleus, given off by the nuclei of certain radioactive elements.

Arete A narrow mountain crest or sharp-edged ridge, commonly present above snowline in rugged mountains sculpted by glaciers. It results from the progressive back-cutting of walls of adjoining cirques.

Bering Land Bridge Unglaciated regions of eastern Siberia, Alaska, the Yukon Territory, and the continental shelf regions between Alaska and Siberia that were above sea level for large parts of the Pleistocene.

Beta (β) particle High-energy electrons given off by radioactive decay.

Bifaced stone tools Stone tools manufactured by striking blows on both sides of a flake of rock to produce a thin, lens-shaped form with sharp edges.

Catchment basin A basin that accumulates the runoff of precipitation from a watershed.

Chitin A nitrogen-containing polysaccharide (carbohydrate) compound that forms the hard outer layer in the skeletons of insects and other invertebrates.

Continental climate The climate over the interior of continents, characterized by extremes of summer warmth and winter cold and by limited precipitation.

Cordilleran Ice Sheet Wisconsin Glaciation ice sheet that covered most of western Canada and southeastern Alaska, extending south into Washington, Idaho, and Montana.

Crustose lichen Flat, disk-shaped lichens that grow close to a rock or other surface.

Cyclotron An electromagnetic machine that accelerates high-energy particles (e.g., protons and electrons) in a circular path; the particles approach the speed of light.

Diatoms One-celled microscopic algae with cell walls reinforced with silica.

Ecological succession The replacement of one kind of biological community by another kind; the progressive changes in a region's flora and fauna that may culminate in a stable, climax community.

Ecosystems Biological communities, including all the component plants, animals, and other organisms, together with the physical environment, forming an interacting system.

Endemic A species whose populations are confined to a certain region, with a relatively restricted distribution.

Exoskeleton The external skeleton of insects.

Flintknapping The process of shaping stone tools from flint or other workable stone by driving flakes from a shaped core piece with a hammer stone or antler.

Fluvial sediments Sediments laid down by running water (streams and rivers of all sizes).

Frost heaving The uneven lifting or upward movement and distortion of surface soils, rocks, and vegetation due to subsurface freezing of water and growth of ice masses, especially ice lenses.

Glacial erratics Boulders gouged out of bedrock by glacial ice, carried along with the ice flow, and eventually dropped as the ice receded.

Glacial moraine A mound or ridge of unsorted glacial debris deposited by glacial ice in a variety of landforms.

Glacial outwash Sediments deposited by streams emanating from the fronts of glaciers.

Glacial terminus The outermost margin of a glacier.

Glacial till Material laid down directly by glacial ice.

Holocene An epoch of the Quaternary Period spanning the interval after the last glaciation (10,000 yr B.P. to recent).

Ice wedge Wedge-shaped ground ice produced in permafrost, measuring from a few millimeters to more than 6 m wide at the top, and extending down from 1 to 30 m. Ice wedges originate by the growth of hoar frost or by the freezing of water in a narrow crack in the permafrost.

Interglacial A long interval between glaciations in which the climate warms to at least the present level.

Interstadial A warm climatic episode *during* a glaciation, marked by a temporary retreat of ice.

Ion An electrically charged atom or group of atoms. An atom with a high affinity for electrons may acquire an electron, thus becoming negatively charged.

Isotope A variety of an element. Isotopes of an element differ from one another in the number of neutrons contained in the atom's nucleus.

Krummholz Growth form of trees, frequently found in stands growing near treeline. The trunks of the trees are shortened and twisted; their branches lie prostrate, near the ground.

Lateral moraine A long, low ridge deposited at or near the side margin of a mountain glacier.

Macrofossils See plant macrofossils.

Marl A sediment deposited in lakes, composed mainly of calcium carbonate and mixed with clay or silt.

Megafaunal mammals Mammal species whose adults weigh more than 40 kg.

Microfossil Fossil too small to be examined without the aid of a microscope.

Microsculpture Microscopic sculpture, including striations, punctures, and meshes, on the hardened surface of plants and animals.

Muskeg A bog in subarctic regions, characterized by an abundance of mosses (usually *Sphagnum*) in standing water and surrounded by shrubs and black spruce.

Nunatak A mountain region thought to have been above the level of glacial ice that may have supported some plant and animal communities through a glaciation.

Pacific coastal forest A temperate zone rain forest that grows in the Pacific Northwest region of North America, from northern California to Kodiak Island, Alaska. It is characterized by Sitka spruce, western red cedar, and western hemlock.

Paleoecology The study of the relationships between ancient organisms and their environments.

Palynology The study of fossil and modern pollen.

Parkland A type of forest in which the trees are widely spaced over a landscape covered with herbaceous vegetation.

Patterned ground Well-defined, more or less symmetrical ground features such as circles, polygons, nets, stripes, and steps, formed mostly by frost action beneath the surface of the soil in arctic, subarctic, and alpine regions.

Periglacial environments Environments at the immediate margins of glaciers and ice sheets, greatly influenced by the cold temperature of the ice.

Photosynthesis The synthesis by plants of carbohydrates from carbon dioxide and water by chlorophyll, using light as energy.

Plant macrofossils Macroscopic remains of ancient plants, including roots, stems, leaves, and fruits.

Plate tectonics The theory that the earth is divided into rigid blocks, called plates, that move relative to one another.

Pleistocene epoch An epoch of the Quaternary Period, spanning the interval from 1.7 million years ago to 10,000 years ago. The Pleistocene is characterized by a series of major glaciations.

Polar deserts High-latitude regions near the north and south poles that receive very little moisture.

Precipitate The separation of a compound from a solution by a chemical or physical change.

Proglacial lake A lake formed just beyond the front margin of a glacier, generally in direct contact with the ice and often dammed by it.

Projectile points Sharp, pointed heads of stone or other material that are attached to a shaft to make a projectile that is thrown or shot as a weapon. These include spearheads, arrowheads, and darts.

Proxy data In Quaternary studies, data from fossil organisms, sediments, ice cores, etc., used to reconstruct past environments; proxy data serve as a substitute for direct measurements of such phenomena as past temperatures, precipitation, and sea level.

Quaternary Period The second period of the Cenozoic era, following the Tertiary and spanning the interval from about 1.7 million years ago to the present.

Stone stripes A type of patterned ground in which sorted, coarse rock debris occurs in bands between wider stripes of finer material.

Stratigraphic column The arrangement of layers of sediments (strata) in geographic position and chronological sequence.

Symbiosis The living together of two or more different species of organisms in such a way as to be mutually beneficial.

Taphonomy The process by which living organisms become preserved in the fossil record; the study of that process.

Terminal moraine The end moraine that marks the farthest advance of a glacier or ice sheet.

Tidewater glacier A glacier that terminates in the sea, often with an ice cliff from which icebergs are discharged.

Treeline The altitudinal limit of tree species in mountain regions and the latitudinal limit of tree species at high latitudes.

Unconsolidated sediments Sediments with particles not cemented together or turned to stone.

Water-lain sediments Sediments that are deposited in water.

Wisconsin Glaciation The last major glaciation in North America, spanning the interval from about 110,000 to 10,000 yr B.P.

INDEX

VERMONT STATE COLLEGES
0 0003 0590581 2